P9-EMK-583

797,885 Books

are available to read at

Forgotten Books

www.ForgottenBooks.com

Forgotten Books' App
Available for mobile, tablet & eReader

Download on the
App Store

ANDROID APP ON
Google play

ISBN 978-1-330-06375-0
PIBN 10016579

This book is a reproduction of an important historical work. Forgotten Books uses
state-of-the-art technology to digitally reconstruct the work, preserving the original format
whilst repairing imperfections present in the aged copy. In rare cases, an imperfection in
the original, such as a blemish or missing page, may be replicated in our edition. We do,
however, repair the vast majority of imperfections successfully; any imperfections that
remain are intentionally left to preserve the state of such historical works.

Forgotten Books is a registered trademark of FB &c Ltd.
Copyright © 2015 FB &c Ltd.
FB &c Ltd, Dalton House, 60 Windsor Avenue, London, SW19 2RR.
Company number 08720141. Registered in England and Wales.

For support please visit www.forgottenbooks.com

1 MONTH OF
FREE
READING

at

www.ForgottenBooks.com

———◇———

By purchasing this book you are eligible for one month membership to ForgottenBooks.com, giving you unlimited access to our entire collection of over 700,000 titles via our web site and mobile apps.

To claim your free month visit: www.forgottenbooks.com/free16579

* Offer is valid for 45 days from date of purchase. Terms and conditions apply.

English
Français
Deutsche
Italiano
Español
Português

www.forgottenbooks.com

Mythology Photography **Fiction**
Fishing Christianity **Art** Cooking
Essays Buddhism Freemasonry
Medicine **Biology** Music **Ancient**
Egypt Evolution Carpentry Physics
Dance Geology **Mathematics** Fitness
Shakespeare **Folklore** Yoga Marketing
Confidence Immortality Biographies
Poetry **Psychology** Witchcraft
Electronics Chemistry History **Law**
Accounting **Philosophy** Anthropology
Alchemy Drama Quantum Mechanics
Atheism Sexual Health **Ancient History**
Entrepreneurship Languages Sport
Paleontology Needlework Islam
Metaphysics Investment Archaeology
Parenting Statistics Criminology
Motivational

SOCIETAL EVOLUTION

A STUDY OF THE EVOLUTIONARY
BASIS OF THE SCIENCE
OF SOCIETY

BY

ALBERT GALLOWAY KELLER

PROFESSOR OF THE SCIENCE OF SOCIETY
IN YALE UNIVERSITY

Cornell University Library

HM106 .K29

Societal evolution;

3 1924 032 560 090

olin

New York
THE MACMILLAN COMPANY
1915

All rights reserved

COPYRIGHT, 1915,

BY THE MACMILLAN COMPANY.

Set up and electrotyped. Published March, 1915.

Norwood Press
J. S. Cushing Co. — Berwick & Smith Co.
Norwood, Mass., U.S.A.

PREFACE

FOR many years the phraseology of evolution has been current. Scientists use it because evolution has come to be the underlying idea of several modern sciences; and less serious writers find their vocabularies colored by what has now become a popular doctrine. Evolution is the fashion, and to affect evolutionary terminology is one method of lending a pseudo-dignity to the trivial. All this is peculiarly marked in writings having to do with sociological subjects.

Naturally the terms originally used by Darwin and his followers have suffered, in passing from hand to hand, a considerable amount of damage. Like coins that have been circulating indiscriminately, they have lost their sharpness of outline and definiteness of superscription. It is almost impossible to discover what some authors who deal with social topics — let alone the host of popular writers and orators — mean by evolution.

In the endeavor, some years ago, to dispel from my mind the vagueness of the evolutionary terminology which had settled there as the result of reading along sociological lines, I went back to Darwin and Huxley — to the mint, as it were — and tried to get their conceptions before me. This was a most useful enterprise, and its results have been incorporated in my thinking and teaching ever since. I have come to believe that any fruitful study of the science of society must rest upon a clear understanding, even though it be but a layman's, of the Darwinian theory.

Then came the question as to the validity of extending Darwinism and its terminology to the life of human society. As to this matter, I have come to believe that the Darwinian factors of variation, selection, transmission, and adaptation are active in the life of societies as in that of organisms. Selection, for example, is none the less selection — not merely *like* natural selection in a vague way — because it is observed in another field and is seen there to act after another mode characteristic of that field.[1] And I have tried to get at the nature of these evolutionary factors as displayed in their societal mode. The outcome of this study takes form in a simple, and in

[1] Cf. pp. 14–16 below.

many ways obvious arrangement of well-known facts; but because the results reached have been of use to me and to my students, in getting and keeping our bearings for the study of the science of society, I publish them, at length, in the hope that some others may derive such advantage from them.

The working out of this line has resulted in an extension upon the conception of the folkways, as developed by my predecessor, Professor Sumner. This conception has always seemed to me to be the link connecting organic and societal evolution, and the extension of which I speak consists in good part in bringing out that point of view, to which Sumner himself gave no special attention.[1] Regarding this essay, as I do, as an extension of Sumner's work, I have taken my quotations, so far as I could, from his writings.

When it came to putting what follows into book form, the choice was presented between accompanying the argument with numbers of cases in illustration, as one does in college lectures, or of securing brevity and compactness by reducing illustration to a minimum. The latter alternative was chosen because, as it

[1] See Note at end of this volume.

seems to me, an essay generalizing about the social order in which we all live is quite likely to suggest its own cases, for or against, to any mature reader who is interested; and then, as a professional teacher, who has gotten to looking at most things from the standpoint of his vocation, I cannot ignore the fact that an expositor generally gains by using his own illustrations. Much is left, in this book, to the reader and the teacher.

I have received much and pointed criticism as I have gone on. For a number of years I have profited by the keen observation and thrusts of my undergraduate students, and by the more mature, though scarcely more helpful comments of graduate scholars. But no other single person has given me such penetrating criticism, accompanied by such sturdy support, as has my colleague in the science, Professor James E. Cutler, of Western Reserve. His attentive reading of my chapters has saved me from many omissions and commissions, especially when it was a matter of making a foray into the field of applied sociology. I herewith recognize my obligations also to Professor Henry P. Fairchild, of Yale, and to Messrs. Charles H. Ward and Julius C. Peter, for especially enlightening

criticisms and suggestions; and to a number of
other friends for reading and passing judgment
upon my manuscript.

ALBERT G. KELLER.

New Haven,
January 11, 1915.

CONTENTS

SOCIETAL EVOLUTION

INTRODUCTION

On a number of counts, the natural scientist is fairly an object of envy to the social scientist. The former deals with things, gets verification through repeated experiment, sets aside unreasonable prejudice by definite proof, attains such certainty as to justify prediction. The latter deals with the elusive human factor, may not deliberately experiment, faces what looks like impregnable prepossession and tradition at every turn, and cannot in general arrive at more than a high degree of probability. Furthermore, natural scientists have a way of climbing by standing on each other's shoulders, whereas social scientists, much like some philosophers and historians, seem to feel that they must each begin at the bottom, and start off by discrediting their forerunners. No doubt some such unlikeness in method lies implicit in the different nature of the two fields;

but it is quite evident which of the two methods is the successful one, and it should be the aim of the social scientist who senses the situation to align himself as much as possible with the more promising procedure.

Such considerations start up before the mind when one is led to reflect upon what evolution means to these two types of scientific workers. To natural science the discoveries of Darwin meant new bearings, a perspective, a unifying principle — and then the inevitable renewal of confidence and burst of eager effort. The idea of evolution showed no kinship with metaphysical speculation. It was no dogma. It is a concrete, proved process, demonstrated in definite terms, and actually verified day by day in the incidents of scientific progress. It is this abundant and continuous verification, not "intuition," which has lent it the force of a major premise. It is not a vague formula which aims to include everything — which is at the same time irrefutable and also useless for the further prosecution of scientific labors — but it is to the intellectually curious both a cheering evidence of the success of scientific method and a tool capable of effective use in further discovery. Natural scientists mean something definite and actual when they use the Darwinian terms — variation, selection, transmission, adaptation — and they are helped to get somewhere by the clearness and significance lent to their scientific terminology as the result of Darwin's life and labors.

It would be strange if a set of conclusions like Darwin's, drawn from such a number of instances over several branches of natural science, should prove applicable only to the field of his special inquiry. The fact is that the doctrine of evolution has so held the interest, hostile or devoted, of the world, that its influence, or at least its terminology, has penetrated into regions quite remote from those in which Darwin labored. In a volume of essays in commemoration of the centenary of the birth of Darwin and of the fiftieth anniversary of the "Origin of Species," [1] notable scholars from many fields united in recognizing their indebtedness to the evolution theory. Here was acknowledged in generous terms its influence, for example, upon the science of language, upon history, and upon the study of religions.

[1] "Darwin and Modern Science," ed. by A. C. Seward, Cambridge, 1909.

The influence of evolutionary reasoning has reached out, then, beyond the field of its nativity; and it is undeniable that it has had its effect upon the social sciences. But no one can say that the evolutionary principle, as commonly viewed by the social scientist, be he economist, political scientist, or sociologist, affords any such orientation, perspective, unification, inspiration, and positive aid as it does in the realm of the natural sciences. It is an idea which he may accept as a sort of philosophy; it lingers in vague form about the horizon; it is invoked, from time to time, in a general and unproductive sort of way. It is a tenet rather than an instrument for the furthering of scientific knowledge, a thing to be discussed by way of formal logic and speculation rather than proved and used over and over again as a scientific verity. It is no tool of research. If social scientists use the terms variation, selection, and so on, they mean nothing definite and actual by them; their use of these terms is vaguely analogical and does not help them to get anywhere.

Perhaps it is worth while to reflect upon the reason for this condition of affairs. The trouble is that the idea of evolution has come to the social sciences through a medium, ostensibly scientific, but really philosophical. It is not Darwinian evolution, in the majority of cases, at all. The common persuasion of the social scientists — who are generally too little versed in natural science — is that evolution means Spencer. In his reading the student of the social sciences becomes acquainted with works which were written under the spell of Spencer; authors vary in their relations with him from the glad discipleship of a Fiske on through to the incensed hostility of the orthodox, but they all unite in revering or assailing him as *the* exponent of evolution. In the natural sciences, on the contrary, not much is heard of Spencer, but evolution means Darwinism. This is very significant of the unlike conceptions of evolution in these two ranges of science.

I have said that social scientists know too little of natural science and its methods. This is a pity, for it is peculiarly needful in the social sciences to keep the feet firmly upon the ground — and the ground is much more likely to be there, and to be solid, where experimentation is, where verification is more positive, substantial, and unescapable, where it is a matter of observable, objective fact rather than of the balancing of ideas. Darwin cannot be too well known to any one who aspires to the name of scientist; and the facts about his life, character, methods, and performances should be at the finger-ends, especially of the man who aspires to apply science to the study of human society. With all his brilliance Spencer is really, for such a person, an untrustworthy guide. He has been passed upon by natural scientists, beginning with the illustrious group of his contemporaries. Huxley, and even Darwin himself, were familiar with both Spencer and his work; and each, while marvelling at his powers, virtually set him aside as a scientist. Says Darwin of Spencer:

"I feel rather mean when I read him: I could bear, and rather enjoy feeling that he was twice as ingenious and clever as myself, but when I feel that he is about a dozen times my superior, even in the master art of wriggling, I feel aggrieved. If he had trained himself to observe more, even if at the expense, by the law of balancement, of some loss of thinking power, he would have been a wonderful man." "I suspect that hereafter he will be looked at as by far the greatest living philosopher in England; perhaps equal to any that have lived." And, again, "I find that my mind is so fixed by the inductive method, that I cannot appreciate deductive reasoning:

I must begin with a good body of facts and not from a principle (in which I always suspect some fallacy) and then as much deduction as you please. This may be very narrow-minded; but the result is that such parts of H. Spencer as I have read with care impress my mind with the idea of his inexhaustible wealth of suggestion, but never convince me; and so I find it with some others. I believe the cause to lie in the frequency with which I have found first-formed theories [to be] erroneous." [1]

Such was the judgment of Darwin. Huxley, a close friend of Spencer, used to poke fun at his "diabolical dialectics"; his classic joke, about Spencer's idea of a tragedy being the destruction of a grand hypothesis by a refractory fact, reflects the essential truth of the matter as respects Spencer's scientific attitude. The fact of it is, that Spencerian evolution is a philosophy. The author called it that. What he sought in the "Synthetic Philosophy" was an inclusive formula. But such a universal is not the aim of science at all. This formula states that: "Evolution is an integration of matter and concomitant dissipation of motion; during which the matter passes from an indefinite, incoherent homogeneity to a definite coherent heterogeneity; and during which the

[1] "Life and Letters of Charles Darwin," ed. by F. Darwin, II, 239, 301, 371.

retained motion undergoes a parallel trans-
formation." [1]   Who can refute this?   Yet who
can use it?   It is doubtless true, yet the working scientist, impatient to be getting forward,
will ask: "If true, what of it?   To what end?"
This formula, in comparison with the Darwinian factors, helps in no way toward scientific discovery; it is not the cloud by day, nor
yet the fire by night.   Rather it is like the
vault of an overcast sky; it covers all things,
but nobody can get his directions from it and
fare forth.   Similarly indicative of the philosophical trend of Spencerian evolution is the title
given by Fiske, Spencer's "American adher-
ent," to his major work: "Cosmic Philosophy."
Such a type of evolution must be unfettered;
it includes

> "den ganzen Kreis der Schöpfung . . .
>   Und wandelt mit bedächt' ger Schnelle
>   Vom Himmel durch die Welt zur Hölle!"

It would also appear that the followers of
Spencer generally come to identify evolution
with progress.   Whatever Spencer himself said
or meant, this seems to have been one of the
chief residua left after immersion in his philosophy of evolution.   This view is not Dar-

[1] "First Principles," § 145.

winian. Darwin speaks of improvement sometimes, and also of retrogression; but both were
forms of adaptation to environment. This
was the basic idea, and it was one which could
be defined objectively, as progress cannot.
Darwin was not much interested in pure inference. For instance, the idea, seized upon
with avidity by those who felt the speculative
spirit move, that all life came from one primordial form rather than from several — the idea
of reduction of everything to a universal,
again — was regarded by Darwin as a matter
of little consequence. He was after something
that stopped far short of a grand dogma or
–ism, and this he found in the idea of adaptation, as Romanes says,[1] "The theory of
Darwin has to do with adaptations in all organic nature, whether they have to do with
species or not." Optimism and pessimism are
alike impertinent when injected into science.
But if evolution meant progress, then it was an
optimistic philosophy, therein corresponding to
the mores of an optimistic age of material prosperity. Hence, in part, the vogue of the
Spencerians.

If the student in the social sciences is familiar

[1] "Darwin and After Darwin," II, 161.

only with the Spencerian idea of evolution, he has a philosophy superior, in many respects, notably clearness, to other systems, and which is not disassociated entirely from science. But he has at his service no such actually tested instrument as has the natural scientist in what he calls evolution. If the former is not circumscribed in his life interests and intellectual curiosity, he must have felt that envy of the natural scientist to which I referred at the outset. Is it possible that he too should have recourse to the same efficient instrument?

In other words, is it not possible to extend Darwinian evolution into the field of the social sciences without sacrificing the essence of its value as exhibited in that of the natural sciences? Writers in the former field have occasionally tried to develop a "Social Darwinism." Much that is useful and suggestive has been brought out as these authors have striven in different ways to show the relation of evolution as they saw it to human society and its life; but, to say the least, the Darwinian idea did not take hold as a result of their ministrations. Haycraft,[1] Ritchie,[2] Kidd,[3] and es-

[1] "Darwinism and Race Progress."
[2] "Darwinism and Politics."   [3] "Social Evolution."

pecially Bagehot,[1] have done service along these lines; Chapin's[2] and Conn's[3] recent essays are, in many ways, useful and admirable. But I do not see that any of these has succeeded in lending to social science anything akin to the practical benefits enjoyed by natural science as the result of the development of Darwinism.

From the side of the natural scientist the efforts to extend evolution into the social field began with Darwin himself, and it was for him an unhappy project. All through the work of Darwin one meets with suggestive hints as to the attitude of an exponent of evolution toward social customs and habitudes, but when he attempted something more systematic, the result was deplorable. It is evident to the reader of the "Descent of Man" that, whatever the cause, for once in his life Darwin has been led to essay waters beyond his depth; Chapters IV and V of the "Descent" do not sound at all like Darwin. Because, in the interest of completeness, he was led to attempt the treatment of man's social qualities and institutions, or for some other reason, Darwin in these chap-

[1] "Physics and Politics."
[2] "Introduction to the Study of Social Evolution."
[3] "Social Heredity and Social Evolution."

ters undertook to discuss such topics as the origin of the moral sentiments. This part of the "Descent" had better have been left unwritten, for, in default of the usual mountains of data from which he was wont to draw his weighty inductions, the great scientist was led to wander hopelessly among the unfamiliar and unfathomable quicksands of the metaphysical and intuitional. In so doing he presents but a sorry aspect to his admirers.

The contributions to the subject from other natural scientists include many isolated, but extremely suggestive passages. Here was where Huxley shone above the rest. In 1900 there was an attempt on the part of three noted natural scientists, Professors Haeckel, Conrad, and Fraas, to elicit a work upon the topic: "Was lernen wir aus den Prinzipien der Descendenz-Theorie in Beziehung auf die innerpolitische Entwickelung und Gesetzgebung der Staaten?" A prize contest was constituted, the chief outcome of which was the publication of the prize essay, by a surgeon, Dr. Schallmayer, entitled, "Vererbung und Auslese im Lebenslauf der Völker; Eine Staatswissenschaftliche Studie auf Grund der neueren Biologie." This book was better than a congerics of suggestions; it was a real attempt to attack certain broad phases of society's life in the light of the theory of evolution as modified by a complete acceptance of Weismannism.

Further, as will be seen in some detail in a later connection, those scientists who, headed by the late veteran Francis Galton, have been studying the problems which surround the subject of eugenics, have set themselves to the solution of certain broad questions involving the whole future welfare of society. A journal of eugenics has been founded, and an *Archiv für Rassenbiologie*. The idea of applying Darwinism to the social sciences has been in the air for some time. However, I cannot see that any of these enterprises really attacks the general issue as I have tried to set it forth. The question I have asked myself is: Can the evolutionary theory, according to Darwin and his followers (I mean such followers as Weismann and De Vries), be carried over into the social domain without losing all or much of the significance it possesses as applied in the field of natural science? It is to this question that the present essay addresses itself.

A point of attack suggested itself some time ago while I was studying the clean-cut little chart of the evolutionary process constructed by Wallace.[1] It is evident that the salient features of Darwinian evolution are variation, heredity, and selection; and that out of the operation of these three comes the fourth, which is really the result of the process — adaptation. These are the simple and concrete terms used all the time by the natural scientists. The thought then occurred to me that if they could be applied to social phenomena, we should have the evolutionary formula carried over into the social sciences. On general considerations the balance of probability would be that a generalization so broad and so universally applicable over a large part of the field of science would not be utterly inapplicable over the whole field. It should at least cast some light upon the processes going on in the rest of the field.

It is clear enough that the idea of taking the issue up as it is done here — of exploring the nature of social variation, social selection, social transmission, and social adaptation — was suggested by the Darwinian system. These terms are in use in sociological writings. The analogy is obvious, and if the idea were merely

[1] In "Natural Selection and Tropical Nature" (London, 1891), p. 166.

to carry it out, a book like this would be super-fluous. But I do not know of any writers who go systematically behind the analogy and try to discover whether or not it covers a broad iden-tity. I shall be charged, doubtless, with "rea-soning from analogy," but I do not feel that the charge is deserved. I find a something in the social field which *is* variation, whether or not it may be *like* what is called variation in the or-ganie field; similarly social selection *is* selec-tion and not merely *like* it. In the social field, also, there is a means of transmission having the essential attributes of heredity in nature; [1] and adaptation occurs in one range of phe-nomena as in the other. These factors have their societal [2] mode as they occur in the life of

---

[1] The fact that certain factors of organic evolution, as, for instance, heredity, are inapplicable on the social domain does not alter the evolutionary character of social phenomena, provided that effective substitutes exist to fill the places of the absent factors. The trajectory of a projectile is still a parabola, whether the motive force be air, gas, muscular reaction, or torsion.

[2] This adjective will be frequently used in what follows. I do not wish to apologize for it in any way, for I think an adjectival form corresponding to the noun "society," and signifying some-thing much more definite than "social," is a present necessity in the development of a science of society. Some of the vagueness chargeable to sociological writing is undoubtedly due to the use of "socius" where "societas" is or should be the real conception in mind.

society, just as they have their organic mode when they appear in the life of organisms.

It would follow that if the salient and determinative processes of organic evolution are repeated in their essence in the life of human society, then we can say that there is such a thing as societal evolution and that we have some definite idea of what it is like.

CHAPTER I

THE HUMAN TYPE OF EVOLUTION

VIEWED as an animal, man shows two striking phenomena: first, dispersal over all the varieties of earthly environment as no other animal is dispersed; and, second, a homogeneity so thoroughgoing that it is impossible to distinguish human species, let alone genera and other wider categories. The widest diversity of environment; the narrowest similarity of structure. At best only sub-species or varieties are definable, and even here there is such disagreement among classifiers as to lend great confusion to the study of ethnology. And the likenesses between men which baffle the classifier are not alone to be found as between human beings separated in space over the earth, but also as between those existing in different epochs of time. The Egyptian records and the diluvial remains unite in proclaiming the absence of radical diversity over the ages. In short, the influence of diverse environment and the passing of much time fail to exhibit man as an animal much modified in visible structure as the result of adaptation.

However, if we survey the groups of mankind scattered up and down the earth, and consequently subject to the most diverse influences of natural environment, we find them differing widely from one another in their cultural response to surrounding physical conditions.[1] But this implies a mental reaction on experience rather than an adaptation of a physical order. The fact is that whatever structural modification there is has been made in the brain, and that the rapidity and success of brain-adaptation has rendered bodily change unnecessary, thus freeing man from the inevitable process as seen among plants and animals, and as in them productive of structural characters which are utilized successfully by botanists and zoölogists in their classifications. Very likely actual structural changes are registered in the brain; of some of these scientists have an inkling, and others may sometime be observ-

[1] "The true difference of mankind," says Count Okuma, as reported in the *New York Times* for October 4, 1914, "is neither in the color of the skin nor in the frame of the body, but is, if any, in the degree of culture itself. It is this difference that distinguishes winner and loser in the struggle for existence."

able with the perfecting of scientific instruments and methods.

At present, however, we are driven to a roundabout method of estimating these brain changes. Mental changes emerge in the form of ideas, and these are capable of materialization or realization. The complicated machine is the materialization of the brain-action of its inventors; it is not mere wood and iron. Every weapon, article of clothing, or other invention (standing as a substitute for structural modification) is in a very literal sense an idea *materialized* or made real; so are all systems and economies in society — in a word, all human institutions. The stage in which a people is, in respect to the quantity and quality of its realized ideas, is called its stage of culture or civilization. The height of a society's civilization thus becomes a measure of its members' success in mental adaptation to the environment in which it lives.

Physical adaptations can be observed, described, explained, and classified along the lines of an evolutionary series. So can mental adaptations, though less directly. If races cannot yet be classified as a result of the study of the cerebrum itself, they can be classified on the criterion of the activity of that organ, as displayed in the sum of materialized or realized ideas. Hence the study of the course of civilization, or of that of one of its factors, is as much a study in evolution as is the investigation of the phases through which general vertebrate structure, or the horse's hoof, proceeds. The mode of evolution is changed; the process goes on.

Thus human evolution diverges from the general course of antecedent evolution. But it is not right to say that all human evolution is mental, nor that mental evolution is absent in the animal. The facts support no such position—to occupy which would be to deny the validity of the whole evolutionary theory, as applied to man, by introducing an impassable period, barren of transitions, at some point;[1] it would be to reaffirm the old-time catastrophic theory. Mental and physical adaptation cannot be exclusive of one another; the two are linked together, as has been intimated. "Brain and mind are reacting upon bone and muscle and

[1] "If," says Darwin ("Origin of Species," p. 174), "it could be demonstrated that any complex organ existed, which could not possibly have been formed by numerous, successive, slight modifications, my theory would absolutely break down." Absence of transitions anywhere thus breaks the course of evolution.

subduing and moulding them to their own mental ends. They are making of the body a fitter expression of the higher mental life. The body is becoming an expression of thought. Muscles of speech and expression are more effective and really more powerful than those of back and legs "[1] — and so are fit criteria for natural selection. But for the present purpose there is more to be gained by passing over the slight observable physical adaptations in man and concentrating attention upon the typical human mode — upon the succession of mental adaptations to be seen in the course of civilization.

Human adaptation is thus typically of a mental order. But it is not of one mind or of a few minds. No civilization (sum or synthesis of mental adaptations) of any importance can be developed by individual or by limited group, in isolation. There must be contact and conflict of ideas, that their variations may be sifted out and a residue of superior adaptations preserved. Civilization is a function of numbers and of the contact of numbers. Human adaptation is therefore social as well as mental.

[1] Tyler, "Man in the Light of Evolution," pp. 38–39.

Further, the fact is sometimes lost sight of, that civilization in its aspect of "power over nature" is really adaptation, not mastery. We get the idea that man does not adapt to environment, but adapts the environment to himself and his needs. But we attain no power over nature till we learn natural laws, to conform and adapt ourselves to them. And then we come to be as dependent upon our adaptations as the bear upon his coat of fur or the woodpecker upon his sharp beak. Our lordship over nature consists in the adroitness with which we learn and conform. Judging by the increase of numbers among the civilized nations and the predominance of man over most other forms of organic life, the human type of brain-adaptation is proved to be a successful one. But its success should not lead us to make a mistake as to its true nature.

If we align, now, the human type of adaptation with that of plants and animals, we find them both representing methods of escaping the fatal sweep of natural selection. Plants and animals exhibit structural modifications, while with man it is his ideas which enable him to oppose or evade the selective forces (starvation, disease, violence) which would reduce numbers or would prevent their increase. Let us see how the two work out as respects growth of numbers. Organic beings exclusive of man tend to increase in numbers up to the limit of the supporting power of the environment; to avoid disaster, when numbers reach the "saturation-point," there is no immediate recourse save flight. The only way in which numbers may increase beyond the supporting power of the environment on any given stage is through the development, under the merciless and protracted sway of natural selection, of some form of structural adaptation. But in the case of man, on the other hand, the entrance of the factor of rapid mental variation, and of selection from among variations, brings it about that both flight and slow painful adaptation are usually eliminated, and an advance in the arts of life (the external projection of mental adaptations) enables him to extract more from the environment; that is, to extend the limit of its supporting power. Further, through the development of mental adaptation along yet another line, man can exercise some form of limitation on numbers which shall slow up their increase toward the limit whither they tend. The peculiar effectiveness of man's adaptation may be seen in the clauses which must be added to the above rule about the growth of numbers, when it is a question of man : Human population tends to increase up to the limit of the supporting power of the environment, *on a given stage of the arts*, and *for a given standard of living* [1] — that is, for a given stage of civilization.

If man's mental reactions are conceived of as uncontrolled by law — as free in the sense of being capricious — there is no object in trying to go on with the present essay, or, indeed, with any scientific treatment of society whatever. But we know from a number of sources, and from the standpoint of *a posteriori* reasoning, that there are laws in accordance with which groups of men are sure to act; human beings are controlled in their actions even where to the casual observer their conduct seems most clearly self-chosen. Of course the reference to "laws" is not to statutes, but to laws that are on the order of "natural law." It is desirable here to get down as near to fundamentals as

---

[1] This formula is a slightly modified statement of Professor Sumner's "Law of Population." He used "land" where I use "environment."

possible, and try to reach some idea of the nature of these social laws.

I wish to take an outstanding example of the above contention. It will be admitted, I suppose, that sex-love is one of the most difficult impulses of man to control. If it could be put under rein, certainly it would seem that other and lesser impulses could be bridled. In fact, it is spoken of often as pursuing its course in the face of any and all attempts to check it. "Love will have its way." Mating is supposed to rest largely upon personal choice, to be mainly a matter of chance. Marriage, we are told, is a lottery. Indeed, the poetically inclined go further and sing in inspired strain about how wrong it is to stand in the way of the divine impulse; and there is a degenerate element abroad that wishes society to surrender all the rights that guarantee its interests in marriage and the family in favor of the free spirit-movings of susceptible and impulsive individuals.

But if we go to the facts, we find that human mating has never been uncontrolled. Among savages, who have no statutes, the taboo proscribes or prescribes so many different kinds of sex-unions that one almost comes to discount altogether the element of personal choice among them. And the taboo is effective, backed as it is by ghost-fear. Among civilized people, also, the sex-relation is subject to regulation which considerably limits the field of personal choice. Aside from laws, which form the crudest and most obvious controls, the sway of convention over this whole matter is more nearly absolute than we are sometimes wont to think. A favorite topic of fiction, which by its very nature strives for effect, and often gets it by making us believe for a time that things are as they are not, is not seldom the satisfactions attending in some especial manner the over-stepping of convention. The prince marries the beggar-girl and all is felicity. Nothing is said of the probable offense caused to the prince or his parents by the beggar-girl's native habits and manners. Because such a situation is diverting to the sense of romance, it proves nothing about the way things are in real life. In a novel-saturated age perhaps such a statement as this is necessary.

What is society's attitude toward a romantic mésalliance on the order of this one? It ostracizes, or, if it does not dare do that, it derides both parties. Its taboo is pitiless, like the primitive one; witness the treatment accorded, in modern England, to a girl who has married her deceased sister's husband, and to the children of that marriage. And as the divergence from code approaches greater proportions, the penalties increase; witness the treatment accorded the Mormons in this country, the adherents of free love, "absolute motherhood," and so on. Note the change that came over the spirit of the reception accorded to Gorky, the Russian writer, when it was discovered that the woman who was travelling with him was not Madame Gorky. Certainly this supreme among the passions is subject to control and direction; and the strongest and most compelling control does not lie in written statute, as any one can see who reflects upon the topic, but in convention. Convention penetrates where statutes are not; and enacted laws are no more than the crystallization of certain of the more tangible manifestations of convention.

If, now, realizing that there are deep-lying forces of societal control, we can get at the nature of the laws that determine convention, or, at least, at the nature of convention, we are approaching the fundamentals of societal laws, and so of societal evolution. It is important here to note the distinction between convention and instinct, for it is significant of the distinction between organic and societal evolution. By convention I mean a piece of behavior that is learned, and by instinct one that is inborn — instinctive. Convention is "second-nature"; instinct, "first-nature." The relation between the two is summarily expressed by saying that convention consistently represses instinct. Brief illustration will present this point. Sneezing is natural enough, but convention taboos it, and conventional people have learned to reduce the paroxysm from an explosion to a delicate murmur. Sneezing is as instinctive or natural a happening as one could well find, but even the most hardened unconventional would try to repress himself if the impulse came at a tense and solemn point of, say, some religious service. If he did not, he would feel the disapproval of those about him. It will be noted that he would have no occasion for repression if he were alone. Robinson Crusoe could be entirely unconventional before Friday came. That which is natural, and so innocent, becomes a fault in society — a fact which reveals the essential truth that convention belongs to societal, not natural evolution.

A child comes into the world, despite the expression "to the manner born," without the beginning of manners. Line upon line and precept upon precept are necessary before he is fit to associate with his fellow-men on terms that do not include the overlooking of his manners. All the conventions taught him, from the time he is made to eat with some implement other than his fingers, until he learns to behave courteously in any sort of surroundings, are repressive of the instincts — all the way from that one which bids him seize and gorge like an animal up to that of an undisciplined selfishness, in the presence of which real courtesy is impossible.

But these conventions are the rules of the game for living in society. No game is worth anything, or pleasurable, or even endurable, which has not its code of conduct; otherwise it runs down into chaos and rough-and-tumble. However, no one made these conventions; they grew up automatically where men lived together. The society, living its life, has imposed, as it were, rules upon its members and a discipline whose absence means anarchy and confusion. The repression of this discipline occasionally leads to revolt, but if the society is to live, it presently settles down to a new code of conventions, which is again enforced with the elemental stress characteristic of society's coercion. For behind any such code, as its ultimate guarantee, is the public opinion of the group-members, dead as well as living. This is the force behind all conventions, and, of course, behind all laws, if they are effective.

Hence if we wish to get down to fundamentals and so arrive at the hidden springs of societal life and evolution, we must seek them in the societal conventions and in public opinion.

In mentioning, in a former connection, several important works touching on evolution in society, nothing was said of a book which develops what seems to a growing body of us to be the connecting idea between organic and societal evolution, or, rather, the fundamental phenomenon of the latter as it diverges from the former into its own phase and mode. This idea is the conception of the "folkways" and "mores" developed by the late William Graham Sumner.[1] No adequate notion of the folkways can be gained apart from the study of this author's analysis, with its supporting compendium of illustra-

[1] In "Folkways, a Study of the Sociological Importance of Usages, Manners, Customs, Mores, and Morals."

tion; but it is yet possible to get before us the conception of the folkways in its broadest lines.

If we recall certain of the salient points of preceding pages, we see that the course of human evolution is the course of civilization. But civilization is a matter of the development of inventions, systems, economies, and so on — of what may be termed, in a broad sense, institutions. In their developed form these institutions are very complex and difficult to handle; but they must have had some simple and informal beginnings and lowest terms (which could be dealt with more readily) in some sort of simple and habitual, unconscious, automatic, unpremeditated reactions upon surroundings. We find such habitudes represented in the conventions, particularly in those of primitive peoples, of which we have been talking. Now here is where the conception of the folkways fits in; as a natural and essential societal form the folkway is analogous to the germ or embryo. It is less derived and more primordial. Folkways represent the lowest terms or matrix of the institutions, whose aggregate, civilization or culture, we have seen to be the external measure and projection of the human type of adaptation to environment. But while it is easy enough to talk vaguely and expansively of convention, it is another thing to possess some definite idea of its origin and nature, and some terminology which shall hold that idea steady amidst the looseness of thought and usage that play about it. Sumner's analysis has defined the conception of these unconsciously arising societal habitudes and conventions, and he has given us an inclusive name for them. "If," says the opening sentence of "Folkways," "we put together all that we have learned from anthropology and ethnography about primitive man and primitive society, we perceive that the first task of life is to live. Men begin with acts, not with thoughts." Thus man's primitive mental operations were more in the nature of nervous reflexes as seen in animals, the only apparent difference between the animal reflex and that shown by nascent man lying in that to which each leads. Here is one of the stock areas of transition characteristic of evolution. If the reflexes are called instinctive, and then later we find the social habitudes or conventions that develop out of them to be repressive of instinct, it must be realized that an advance of socialization has intervened. In evolution no sharp line of distinction can be laid down at the outset, nor yet for some time after lines of development have begun to swerve one from the other.

Men begin, then, "with acts, not with thoughts. Every moment brings necessities which must be satisfied at once." [1] These necessities are those of adaptation, and the sign of their presence is pain; and upon the experience of pain there results an attempt—a sort of reflex action, under some guiding instinct come down from man's animal-ancestry, perhaps — to satisfy the need. In the absence of any experience of the relation of means to ends, while the old need may be met with some precision, new needs call forth but clumsy and floundering efforts for their satisfaction. "The method is that of trial and failure, which produces repeated pain, loss, and disappointments." Here is certainly a case of variation, and the several variations could scarcely be of equal

[1] What is here said about the folkways is a paraphrase, with occasional direct quotation, from Sumner's book (chiefly from chs. I, II, III, V, XV, XX). I have therefore thought it unnecessary to equip the text with series of references, or to disfigure it with many quotations, or to reproduce many illustrations; but rather to make my general acknowledgment in this place once and for all.

D

value in the struggle to live. Hence there must come about a selection of these variations, since early man is subjected, clearly enough, to the same tendency to outrun the supporting power of his environment, and so to the struggle for existence, as are other organic beings. "Pleasure and pain, on the one side and the other, were the rude constraints which defined the line on which efforts must proceed. The ability to distinguish between pleasure and pain is the only psychical power which is to be assumed. Thus ways of doing things were selected, which were expedient. They answered the purpose better than other ways, or with less toil and pain. Along the course on which efforts were compelled to go, habit, routine, and skill were developed."

It must be recalled, at this point, that human evolution is not an individual matter, but one of societies. It is easy to show, even from instances derived from the animal world, that association is in many cases an important advantage in the pursuit of the struggle for existence, and so becomes a basis of selection. Doubtless the earliest groups of men were very small, but, judging from the most primitive types of *homo* that have been found, association in societies seems to have characterized the race. This is where natural selection seems to have prepared the way for a later form of evolution. If, however, the struggle for existence was carried on in groups, each member could profit by the others' experience; "hence there was a concurrence towards that which proved to be most expedient. All at last adopted the same way for the same purpose; hence the ways turned into customs and became mass phenomena. . . . The operation by which folkways are produced consists in the frequent repetition of petty acts, often by great numbers acting in concert or, at least, acting in the same way when face to face with the same need. The immediate motive is interest. It produces habit in the individual and custom in the group. It is, therefore, in the highest degree original and primitive. By habit and custom it exerts a strain on every individual within its range; therefore it rises to a societal force to which great classes of societal phenomena are due."

"It is of the first importance to notice that, from the first acts by which men try to satisfy needs, each act stands by itself, and looks no further than the immediate satisfaction. From recurrent needs arise habits for the individual and customs for the group, but these results are consequences which were never conscious, and never foreseen or intended. They are not noticed until they have long existed, and it is still longer before they are appreciated. Another long time must pass, and a higher stage of mental development must be reached, before they can be used as a basis from which to deduce rules for meeting, in the future, problems whose pressure can be foreseen. The folkways, therefore, are not creations of human purpose and wit. They are like products of natural force which men unconsciously set in operation, or they are like the instinctive ways of animals which are developed out of experience, which reach a final form of maximum adaptation to an interest, which are handed down by tradition and admit of no exception or variation, yet change to meet new conditions, still within the same limited methods, and without rational reflection or purpose. From this it results that all the life of human beings, in all ages and stages of culture, is primarily controlled by a vast mass of folkways handed down from the earliest existence of the race, having the nature of the ways of other animals, only the topmost layers of which are subject to control, and have been somewhat modified by human philosophy, ethics, and religion, or by other acts of intelligent reflection." Life is full of customs and small ceremonial usages; all men are subject to them and act under them, with only a little wider margin of voluntary variation than that shown by the savage. ╱

"No objection can lie against this postulate about the way in which folkways began, on account of the element of inference in it. . . . We go up the stream of history to the utmost point for which we have evidence of its course. Then we are forced to reach out into the darkness upon the line of direction marked by the remotest course of the historic stream. This is the way in which we have to act in regard to the origin of capital, language, the family, the state, religion, and rights. We never can hope to see the beginning of any one of these things. Use and wont are products and results. They had antecedents. We never can find or see the first number of the series. It is only by analysis and inference that we can form any conception of the 'beginning,' which we are always so eager to find. . . . The origin of primitive customs is always lost in mystery, because

when the action begins the men are never conscious of historical action, or of the historical
importance of what they are doing. When
they become conscious of the historical importance of their acts, the origin is already far
behind."

If this paraphrase and these quotations have
given the reader the fundamental idea of the
folkways, he has before him the conception of a
stage of societal development lying beyond the
purely organic stage, while plainly allied to it
in most essentials; and yet falling short of the
stage of societal development characterized by
compact institutions and a conscious policy.
For the present purpose it is not deemed essential to go back and seek to work out the
connection with organic nature through the
attempt to trace the folkways back into the
unconscious; nor is it conceived to be of consequence that we should know, if we could, at
just what point the characteristically human
mental reactions, as distinguished from "instincts," came first into play. What we wish
to do is to see whether the essential ideas of
Darwinian evolution can be applied to the
course of civilization; and since we can scarcely
fail to admit that the early stretches of the

course of civilization are to be found in the folkways,[1] we should get the more fundamentally at the issue before us by trying the Darwinian ideas, at first, at least, upon the folkways. Here we are working in lower terms, which is always an advantage in attacking a complex and difficult matter.

Summing up at this point the several considerations which precede: we see that man possesses in the brain a sort of specialized adapting organ which relieves the rest of the body from the necessity of structural adaptation; that the human mode of adaptation is thus mental, and that it is also social; that the measure of human adaptation is the degree of civilization attained; that the story of human evolution thus becomes the story of the evolution of civilization in human society; and that the law of population must receive characteristic modifications when it is applied to man. The brain becomes *the* organ of adaptation. Looked at in one way it secures adaptation for man by transforming his environment; but in a broader and truer sense, by learning the

[1] If the sociologist cannot vary conditions and so study phenomena under diverse and selected conditions, he can yet go where, as among primitive people, the phenomena exist under diverse, if not chosen conditions, and study them there.

laws of nature and devising ways of conforma-
tion to them. But the details of this new
phase and mode of adaptation are no longer
matters of biology; the reactions of the individual are cerebral and psychical. However,
these reactions do not remain individual and
isolated, but, in ways indicated in this chapter,
they become societal and so fall into the domain
of sociological study. So soon as concurrence
of reactions takes place a new evolutionary
mode arises. This is a stage removed from
organic evolution, but the connection is unbroken. Inquiring now as to the fundamental
factors and processes of societal evolution, we
have found human actions to be controlled,
not hit-or-miss, and to be controlled in last
analysis by convention, which is repressive of
"instinct" and which is guaranteed by public
opinion; and we have followed Sumner's
analysis and characterization of the folkways.

Since the folkways of a society include all
its forms of mental reaction on environment —
are, in fact, the irreducible elements on any
stage of civilization — it is now possible to
modify somewhat the formula with which we
started:[1] The measure of the characteristic

[1] P. 19 above.

type of human adaptation is the body of its folkways and mores.

The folkways are the simplest and most fundamental phenomena of societal life. They form the germ and matrix of all human institutions.[1] It is to them, then, that we can best address ourselves in inquiring as to the application of the Darwinian factors to the life of society. What we wish to consider is this: whether these factors obtain in the folkways and their derivatives, and whether, if they do, the final result is, in the one case as in the other, adaptation. If it can be shown that the folkways are adaptive by way of the activity of factors of the order of those operative in organic evolution — factors which actually operate in the societal domain in the same way that those do in the organic world — then there is no reason to refuse to extend the Darwinian theory, in its new mode, into the "superorganic domain." But in treating of evolution in the mores, one must not forget that he is operating, as it were, in a sort of secondary intension; he must always be ready to refer back to the underlying

---

[1] For a vigorous scientific treatment of law as a development out of the folkways, see Corbin, A. L., "The Law and the Judges," in the *Yale Review* for January, 1914.

organic processes, or he gets his feet off the earth and risks losing himself in vague speculation. Grounds for this caution will appear as we go on.

CHAPTER II

VARIATION

VARIATION is the opposite of uniformity and monotony and is the basis for all need of classification. If there were no variation in nature, there could be no such multitude of forms, shading into one another by almost imperceptible gradation, as the naturalists tell us there is; similarly, if there were no variation in the folkways, human customs and institutions could show no such endless diversity of detail as we see about us. These are generalities. To demonstrate the existence of variation in the folkways would be a matter of rehearsing long series of instances, most of which would doubtless be more or less familiar to the reader, and many of which are to be found in the collections of cases, under various topics, in "Folkways." It does not seem necessary to assemble such exhaustive detail. Any one can see, on brief reflection, that no two human groups — whether family, club, sect, secret society, township, state, or nation — have the same code of mores. Even the code of personal conduct shows its variations and, by virtue of selection among them, is altered from time to time.

Variation in the folkways is practically self-evident. If the first reactions on environment characteristic of men were clumsy, groping, and non-purposeful, then variation must have been there at the outset. The dream of a medicine-man has often set afloat a new societal variation. But random and inconsequent variation is not confined to primitive man — Joan of Arc too had her vision; it can be observed in every age, especially in the case of folkways not subject to conclusive test, as, for instance, in fashion. Variation comes to be limited after selection has operated for a time. The process is this: a considerable degree of concurrence having come about with regard to certain folkways, these are taken to be conducive to group welfare, and presently receive social and religious sanction and become mores.[1]

[1] "When the elements of truth and right are developed into doctrines of welfare, the folkways are raised to another plane. They then become capable of producing inferences, developing into new forms, and extending their constructive influence over men and society. Then we call them mores. The mores are

Then variations from them are, under this conviction and sanction, prejudged as inexpedient, and so, to recall a phrase quoted above, there is laid down a "course on which efforts were compelled to go." This would be like building two lofty walls on either side of a growing tree; the outspreading of its branches must then take place within the narrowed space. So variations could appear only within the "course" left open. In our own societies the folkways governing diet no longer permit of variations in the direction of cannibalism; and those of war are supposed to exclude all variations in the direction of using poisons, employing certain types of treachery, and so on. But within the limits laid down, variation still goes on: new ideas about eating characterize the folkways of the time, and more new devices for killing men abound among civilized people than ever the savage could dream of.

It would be easy to immerse this subject of

the folkways, including the philosophical and ethical generalizations as to societal welfare which are suggested by them, and inherent in them, as they grow." Sumner, "Folkways," § 34. It will not be practicable always to maintain a clean-cut distinction between folkways and mores — indeed, Sumner himself does not do so — but that there is "another plane" should be kept in mind.

variation in analogies: to talk of mutations in the mores; to refer to the amalgamation of groups and the consequent development of new combinations in the code as social amphimixis, thus suggesting an explanation of variation along the lines laid down by Weismann. This sort of procedure is entertaining and might be suggestive; but it is dangerous. The fact of variation in the folkways is all that needs to be established here. No doubt when two or more codes are brought into contact by the compounding of groups, there appears to ensue an activity of variation in the mores that suggests fecundation of some sort. Perhaps there is really something more in such a case than mere inter-transmission of folkways between the groups thus brought into contact. But if there is, it is safer with our present knowledge to ignore so vague a quantity and to treat the whole matter objectively under the topics of selection and of transmission by contagion.[1]

Conscious variation within lines allowed by the folkways is now called experimentation. It is supposed to be scientific, that is, carefully directed on the basis of law and of knowledge of what has been done before; but it is not

[1] Pp. 73 ff. and 232 ff. below.

seldom as clumsy and floundering, and as unconscious of the previous history of the race,
as are those wild procedures of the savage to
which we would deny the name of experimentation. The rich and powerful, those who have
leisure or means to experiment, can set new
variations afloat; and so can the sensational
demagogue, living on his wits, and thriving by
reason of the fertility of his inventive and per-
suasive powers. Such have fathered many of
the "social experiments" of the modern age.
But, however "happy ideas" and "projects"
may be viewed, it is clear that variations in
the folkways, conscious or otherwise, have
always been occurring, sometimes perishing at
once, sometimes attaining a small following
among the concurrent, and again uniting huge
factions. The collections in the Patent Office
form a museum of variations in the purely
mechanical field, the courts and the newspapers
offer numerous cases in the realm of marriage
and the family,[1] the records of administrative

---

[1] "Every possible experiment [in the way of sex relations and
marriage], compatible with the duration of savage or barbarous
societies, has been tried, or is still practised, amongst various races,
without the least thought of the moral ideas generally prevailing
in Europe, and which our metaphysicians proclaim as innate and
necessary." Letourneau, "Evolution of Marriage," p. 344.

bodies of importance or insignificance provide examples of variation in the political mores, and the history of sects and creeds teems with instances of the same process in the matter of religious system and ritual. Fads and fancies are the froth of all this variation.[1] Similar diversities can be shown in the manners and customs of human groups upon all stages of civilization; there seems to be no danger of any stiffening into monotony. *Quot homines, tot sententiae.* At any stage there are plenty of budding variations in the folkways to select from.

These human variations are not periodic, depending upon birth-periods of new organisms; they take place in rapid and sometimes ap-

---

[1] A ridiculous case of fad is reported by the *New York Times* for January 25, 1914. "Professor Bergson's lectures on philosophy at the Collége de France are becoming one of the social functions of Paris, with the result of an amusing conflict in the lecture hall between the society dilettanti for whom philosophy is the present mode and the students who attend the lectures as part of their course for a degree. Before the time fixed for a lecture every seat is usually taken by society women, who, because the university is a national one, cannot be excluded. The bona-fide students cannot come early on account of other lectures. Already there have been several scuffles in consequence. Maurice Croiset, the head of the college, has made fruitless appeals to the smart set, which refuses to be deprived of its new diversion. A protest is now being made in the press."

parently in inconsequent fashion; ideas tumble over each other in the face of the need which evokes them. But selection cannot, as will be seen, keep pace with the production of that upon which it is to operate, and so the unfittest variations are not eliminated with the expedition characteristic in nature. Further, new variations depend upon older ones which may not have been tested as yet; plenty of ideas may gather about the elaboration of unverified principles or happy thoughts. The mores show a capability for producing inferences. So that when the original variations come up for judgment, the case is already confused by the existence of numerous corollaries, all of which cannot well be eliminated along with their supporting principle. For instance, if protectionism might be taken as a variation in social policy, there are not a few who reject the principle, but fear to select it out for extinction because of the multitude of societal structures which have developed in dependence upon it.[1] It is plain that the evolution of

---

[1] Illustrations appear in current literature; the case of the size of bricks is adverted to by the *Philadelphia Inquirer* (toward end of August, 1911). "If bricks were made larger it would save a great deal of time and labor in building, said a contractor, but

E

society and of social forms is an extremely complex affair and should be approached in anything but a jaunty manner. Something is gained if this point alone is apprehended.

Viewing this topic of variation in the large, it should still be kept in mind that in societal evolution as in organic there can be no comprehensive view of variation over long epochs by reason of the extinction of transitional, non-advantageous variants; and in the one case as in the other many a mean between extremes has been preserved, if at all, only under the protection of isolation or in the form of a fossil or survival. Such fossil variations, persisting occasionally in isolation, are cannibalism and human sacrifice, group-marriage and polyandry, communal property and inheritance by nephews

the standard has been set and any change would be attended by considerable inconveniences. In England when bricks were first made, and up to sixty or seventy years ago, there was a tax on bricks and in order to evade it the bricks were made of larger and larger sizes.

"These were used for cellars and other concealed places. To stop this fraud, an act was passed in the reign of George III fixing the legal size of bricks. Early in Queen Victoria's reign the tax was taken off and the bricks may now be legally made of any size whatever. But any change from the standard size would bring about great inconvenience. All calculations are made for building on this standard size, and the London and other building acts have practically fixed it."

— outreachings in the effort to live which were not destined to persist.

The case of societal variation reduces ultimately, then, to the mental reaction of individuals. These, unconsciously, and later to some extent consciously, throw out a series of tentatives under the stimulus of need. Certain of these tentatives cancel out at once or otherwise disappear, while others are concurred in and become characteristic of a group. They are then the folkways of that group, and when they become the object of group approval and so become the embodiment of its prosperity-policy, they become its mores. Here is where they enter into the field of societal evolution and invite the interest of the sociologist; for they are now social phenomena as distinguished from individual phenomena. It must not be forgotten that they probably go back to physical change in the individual brain, and so root in organic processes and organic evolution, and in the resultant "race-character" or temperament. But, having made this connection clear, and having thus laid yet another claim on behalf of societal evolution to an origin in organic evolution, we need no longer concern ourselves with the individual as with a necessary factor in our argument. The origin of and the variation in the mores may go back to the individual, but they themselves are group-phenomena and their selection, transmission, and adaptation are characteristically societal, not individual — just as civilization is a societal product and incapable of advance under isolation.

CHAPTER III

FROM the preceding chapter we carry away the fact of variation in the ways and between the codes of groups, based upon concurrences along certain lines struck out upon by individuals. But the very fact of concurrence implies selection already in the field; thus selection seems to exist on the stage immediately preceding the formation of the folkway. This, however, does not concern us except as it exhibits one of the transition forms which link together organic and societal evolution and which in so far justify the conviction that the evolutionary system does not stop short of man and human society. Individual tentatives are the object of selection; and concurrence in them, if they are apt, means the preservation of some at the expense of the elimination of others. This must be so even in a society of animals, for they fall into group-ways also; in fact, there are places enough where it would be difficult to draw a sharp distinction between some animal ways and some folkways.

But, for the purpose in hand, concurrence is supposed to have taken place, and we have before us over the earth numerous aggregations of folkways characterizing various groups, smaller and larger. If we take the family as the smallest group, we may say that no two families are likely to have identical ways; and if we take the nation as the largest group, that there are no two nations whose customs and habitudes will exactly correspond. Here is evidence, retroactive, as it were, for variation in the mores. And here is, in any case, a good ground for selection; on the organic stage, such a condition would demand and evoke selection without delay, whether such selection were artificial, at the hand of the breeder, or natural in the most elemental sense.

It is a commonplace of history that customs and institutions arise, attain strength and vogue, finally to yield to other ways of procedure and to disappear. For the explanation of this rise and fall we do not look to the individual or to the limited group of individuals, except as he or they stand forth as the leading agency or exponent of the societal change. This is the only way in which the sociologist can view the "great man in history"; he may sign an Emancipation Proclamation, but he does it only as a delegate or representative. For the causes of alteration in the code we look to more massive factors which have to do with the life of the society as a whole, as, for instance, economic change or the rise of a religion. No great selection of the folkways can take place apart from the swinging of the body of the society; there is a "social movement" and the result is a "new social order," that is, an altered body of mores. It is the society which performs this societal selection, just as it is "nature" which performs natural selection. By nature, whose personification is so difficult to avoid, we mean here, with Darwin,[1] "only the aggregate action and product of many natural laws, and by laws the sequence of events as ascertained by us"; similarly by society, or the social order, we mean here only the aggregate action and product of many societal laws, and by societal laws the sequence of societal events as ascertained by us. There is, as will be seen, something about societal selection which is after the model of natural selection, massive and elemental.

[1] "Origin of Species," p. 75.

To bring about any selection at all there is need of conflict as between variations. In the case of the folkways this contest, plainly enough, cannot take place as between ideas, though we may figuratively represent it as so doing; as, for example, when we speak of the conflict between conservatism and radicalism, religion and science, and so on. The real struggle is between the adherents or exponents of the mores in question, and it is through the issue of that struggle that a given set of mores is carried forward toward universality or consigned to oblivion.

That such a struggle between groups characterized by different codes of mores shall never cease is a matter which is settled in the order of the universe. The struggle for existence — the securing of a food supply — is in itself sufficient to assure conflict between organic beings of all grades. There will always be conflict where there are wants and insufficient means to satisfy all. And it is provided in human nature that wants multiply and diversify as they are about to be satisfied. And when the habit of association has been evolved, then the struggle is group-wise. Driven by their interests, groups of all sizes, from the race-group down to the smallest, are always in conflict of some kind with their competitors. Their relations are those of latent hostility, or at best of antagonistic coöperation.[1] The reader may turn to Gumplowicz for an eloquent exposition of this "Rassenkampf." Wherever interests coincide in their demands there is ground for competition and conflict; and the main interests of all human groups are identical: to live, and then to attain a higher standard of living.

But there exists in the mores themselves a ground for antagonism due to differences in code, which often cloaks over a more elemental ground. The mores of a group are not seldom offensive to another group; and where they are not that, they are contemptible or laughable. It would appear that certain groups brought into contact with others find the ways of the latter so intolerable that they may deliberately and consciously set out to get rid of them. We must pause here in our discussion of societal selection to look into this matter. It has been seen that the code of mores of a group becomes its prosperity-policy; the mores in which the group has concurred are thought to be conducive to societal welfare, and departure from

[1] Sumner, "Folkways," § 21.

them is regarded as perilous. Thus they become "uniform, universal in a group, imperative, and invariable," growing, as time goes on,
"more and more arbitrary, positive, and imperative." [1] They are thought of as the code
of a superior group, and this involves their
comparison with the codes of other groups, to
the disadvantage of the latter. The Kubus
of Sumatra believe illnesses arise from relations
with people outside the tribal group.[2] This
group-egotism, which, among other things,
causes so many tribes to denominate themselves "Men," as distinguished from the rest
of the world, who do not measure up to that
exalted title, is called ethnocentrism. The
reason why the rest fall short of "us" is because
of their ways far more than for any other, for
example, any physical peculiarity. Ethnocentrism is thus a specifically human sentiment. It
enters to strengthen the local code of mores as
the distinguishing character of the group, and
to promote intolerance and hostility as respects
the ways of others. "Each group thinks its
own folkways the only right ones, and if it
observes that other groups have other folkways,

---

[1] Sumner, "Folkways," § 21.
[2] "Die Kubus von Sumatra," in *Globus*, XXVI, 46.

these excite its scorn. Opprobrious epithets are derived from the differences. 'Pig-eater,' 'cow-eater,' 'uncircumcised,' 'jabberers,' are epithets of contempt and abomination." A galaxy of such terms could be gathered in our own society and time, as, *e.g.*, bog-trotter, dago, sheeny, griner, hunkie, bohunk, guinea, wapp. These and other terms have been invented to mark the exponents of uncongenial mores, racial, national, or sectional. Thus "ethnocentrism leads a people to exaggerate and intensify everything in their own folkways which is peculiar and which differentiates them from others. It therefore strengthens the mores." [1]

It is to be noted that the differences which catch the eye and are thus held up to contempt are often entirely inessential. Diversity in language is prominent among these; ignorant people take the attitude, so graphically portrayed in "Huckleberry Finn," that a human being should talk in the way human beings were meant to talk, *i.e.*, as "we" do. Again, it is what the other people eat that arouses our contempt and even ire. Greek and American Indian alike despised the "Raw-eaters" (" ὠμόφαγοι," "Eskimantsic"); and the British sailor

<hr>

[1] Sumner, "Folkways," § 15.

hastened to smite the snail-eating Johnny Crapaud. Such judgments, often totally irrational, as to the undesirability of others' mores, have contributed not a little, with the proper opportunities, to the attempt to eradicate both mores and men.

Ethnocentrism as a factor securing loyalty to code is supported by another agency even stronger, and yet farther removed from rationality: religion. In their origin, as foregoing passages have indicated, the mores arise entirely apart from animistic and daimonistic beliefs, and probably long before these. Having become the prosperity-policy of the society, they are inculcated by those in the society who possess the chief influence, that is, by the old; for the elders, the repositories of tradition, those who remember, have ever been the guardians of the mores. Then the aged die and become beings of a superior power; however, in accordance with the prevailing views of the future life, they are supposed still to cling to the ways which they approved while among the living, and still to be ready to lend their power to suppress departures from those ways. But their power is now infinitely greater than it was in life, however redoubtable it might then have been; hence the supreme sanction, that of ghost-fear, is added to the mores, thus securing their persistence and their arbitrary quality. "If asked why they act in a certain way in certain cases, primitive people always answer that it is because they and their ancestors always have done so. . . . The ghosts of ancestors would be angry if the living should change the ancient folkways." [1] This attitude toward the code reënforces the ethnocentric element which, as has been seen, is itself of great strength, and engenders towards other codes a hostility whose driving agency is not alone group-vanity but also the fanatic fear and devotion connected with relations with the supernatural. The religious sanction of the mores is so powerful and engrossing that in the case of group-conflicts the divergence of the mores which lay at the root of hostility has been lost sight of by the actors and even by historians, in the collision of the respective sanctions of these mores. Not infrequently in missionary enterprise of the past the essential difference between a "lower" and a "higher" people, between "heathen" and others, has been taken to lie in the type of sanction accorded the

[1] Sumner, "Folkways," § 1.

mores rather than in the mores themselves. Attempts have been made to alter the sanction of the mores, when the mores themselves form the only real point of attack.

Both ethnocentrism and religious preoccupation, as forming an attitude of mind making for hostility to other codes, are essentially irrational. It is by this quality that they show their primeval character; they have been in and of the mores since remote times — though it must be added of religion that it takes a developed and systematized form to show what thoroughgoing intolerance is.

Perennially fanned by elemental factors of this order, group-conflict has never ceased, and it is unthinkable that it should cease while there are wants toward whose satisfaction men must strive, but for whose universal satisfaction there is an insufficiency of means in the world. Thus are the exponents of diverging codes of mores led into unceasing conflict with each other. This struggle may take place in diverse fields — military, industrial, political — and it is of various degrees of intensity. War, resulting in the annihilation of one group by another, is the primordial agency of selection in the mores, and probably the most efficient that has ever existed. It is after the order of natural selection. But war presently issues in subjection rather than destruction; and subjection is not of a single type, entailing always the same degree of surrender of mores on the part of the subjugated. A conquest may be succeeded by enslavement, but this generally runs out into milder forms of subjection, such as exclusion from certain social privileges and benefits. We shall work down through this series so far as it may seem profitable, by way of introduction to the later, modern form of selection.

Before a decisive test of conflict comes, the variations in the mores of two groups may have decided its issue in advance. There are such things as harmful mores [1] — codes which have developed or persisted in the absence of a decisive competition and test. For instance, among some of the Shingu Indian tribes, which were apparently in decline, the doctrine prevailed that the old men should have the young wives and the young men the older.[2] It does not appear that anything short of conflict can

[1] Sumner, "Folkways," §§ 28, 29, 65.
[2] Von den Steinen, "Shingu Tribes," in *Berlin Mus.* for 1888, p. 331.

select out such variations; it is sometimes said that "invention is the motive power and environment the selective factor," but if environment means merely physical nature, its selective power is in any case very slow-working. Where, however, group-conflict has been persistent, the path of history is strewn with discarded codes. Mores which have been superseded in the evolution of conflicting races, *e.g.*, polyandry, the mother-family, cannibalism, incest, are still found persisting under isolation; or, as will be seen, reappear under isolation. If certain mores physically or numerically weaken a group, or impair its organization, rendering it ever so little inferior to other groups, this fact, which may remain long concealed under isolation, is revealed at once when conflict arises. Such revelation has been, in history, one of the characteristic functions of war.

It is probable that the salient and characteristic features of all prevailing codes of mores represent the residue from such rude tests, involving the virtual extirpation of the losers, with the resultant elimination of many variant codes. Milder forms of selection seem never to have gotten the results; but the war selection has been scarcely less rigorous and pitiless than that which has produced the advancing adaptation of plant and animal forms.

Selection in the mores through the death of their adherents is not confined to inter-group relations; it occurs within peaceful societies. Religious mores have been uprooted through the eradication of those sects who held to them; and divergences from the accepted codes, if they are wide enough, are regarded even now as deserving extinction through the capital punishment of those who practice them.

The destiny of such non-conforming elements approaches that of the "unfit" in nature; they are removed from the society through isolation behind the bars or in more drastic manner. If such rebels against the code are numerous enough, there may come about in the society an upheaval of civil strife and a reversion to primordial methods of selection. It must be realized that at any stage of the selective process violence and bloodshed may reënter. In fact, all forms of compulsion within the group are reducible to violence and killing in the end. Behind the power of the courts, the police, and the rest of the regulative organization of a society lies the force of the army, and, ultimately, of a predominant section of the

F

group-members, mobilized for mortal combat in civil war. It is salutary to keep this consideration in mind as a background, as one moves into modified or derived forms of selection. And upon such an upheaval or revolution there attends a selection so rigorous as to result sometimes in a "new dispensation." A society may be so profoundly stirred that the results reach far beyond the settlement of the issue over which conflict has arisen; for all the various parts of the structure of society are intimately connected one with the other, and the "strain toward consistency"[1] in the mores brings it about that no considerable change can take place in one set of societal forms without modifying the whole structure. Let one consider the elimination of measures and men and the change in the society at large that followed upon the American Civil War.

Where group conflict is to the death, selection among the incipient folkways is speedy and sure; and the occasional catastrophe or ferocious conflict of later ages reveals the presence of inexorable, elemental forces beneath all human life. But as the mores develop and civilization advances, the nature of the test becomes

[1] Pp. 141 ff. below; cf. Sumner, " Folkways," § 5.

decreasingly severe, eliminative, and decisive. Not only are the sanctions of life and death displaced, but even where they still remain, their incidence is often determined by the mind of man rather than by the elemental forces of nature. Let us consider the effect of the increase of civilization on the nature of this selection.

It is clear that the apparatus of civilization is calculated to shield men from the action of natural selection; to interpose a bulwark between them and the physical environment; to ease up the conditions of the struggle for existence. The highest civilization develops the human type farthest removed from the "nature-man"; it thwarts the action of natural selection most thoroughly, that is, it shows the greatest "power over nature." Whole series of factors by which natural selection operates in the non-human organic world are eliminated, at least in large degree, viz., famine, cold, physical conflict between individuals, unrestricted birth rate, and so on. The environment is artificialized, physical competition is lessened, the conditions of strife are so altered as to remove the advantage which the "fit" under nature would enjoy. Or, taking it in the reverse direction, as we go down the scale of civilization, these impediments to the action of natural selection drop progressively away till we get to the primitive people who live practically under its sway. The effectiveness of natural selection varies inversely as the height of civilization. Natural selection develops the man-animal with specialized mental adaptability; this quality (given association, which is also a product of natural selection) works out into the folkways and mores; but these latter operate to exempt man from natural forms of selection, by developing a societal selection which spares those who would not be spared under nature (counterselection).[1] It is possible to take one point of view according to which, starting with natural selection, we seem to proceed by inevitable transitions straight to its opposite. This will be shown, I think, to be a partial view and a one-sided position; but a realization of what truth there is in it helps to sharpen the conception of the change in mode shown by selection, as it accompanies the advance of civilization.

It is necessary to conceive of this alteration of mode in selection as clearly as may be. It

[1] Cf. chap. VI below.

has softened as the mores have come to include elements, like the high valuation of human life, which were not in them in earlier stages. Then these mores, possessing "the authority of facts," have "become capable of producing inferences" resulting in developments of the code which, it would seem, must ultimately be selected away. For instance, along with the advance of civilization there has developed a sentimentality which is, in many of its manifestations, pure weak-mindedness. Ethical philosophers, removed from contact with the facts of life, have evolved, as inferences, a set of ideals and dogmas about human relations which they have succeeded in putting into the minds of the emotional and susceptible. The latter are prosperous enough to pay without much self-sacrifice for the sensation of rectitude which they feel in living in accord with these noble sentiments; or, rather, they are able to make those pay who do not agree, or who are too fully occupied in the struggle for a living to have much time for sentimentalizing. These arbiters of ethics and their following raise horrified outcries at the imposition of the death penalty, at the public lashing of a wife-beater, at the insistence upon an adequate discipline in schools. They make the home a hothouse instead of a toughening training school for life, turning loose upon society undisciplined products prone to disregard the rights of their fellows in society as they have overridden under indulgence the rights of their fellows in the home. Sentimentalists, warm of heart, but soft of head, petition complaisant executives to let loose upon society the wolves that have been trapped and should have been eliminated once for all; to set the scotched snakes free again. The pseudo-heroic and pathetic aspects of the life of a black-hearted criminal are rehearsed until he seems to be a martyr, and the just judge who condemns him a persecutor and a brute. All of which is done by volatile spirits under the illusion that they are thereby conserving the delicacy of the "ethical sense," or what not, instead of proving recreant to plain duties as members and supporters of civilized society.

What sort of selection is this? Seen in its more pronounced forms it seems to reverse the findings of science and of sense in favor of an aimless emotionalism. One is ready to believe that there is in modern civilization nothing analogous to the elimination of what is unfit. We should not lose sight, however, of the fact that the natural processes underlie all others and cannot be set aside. What human beings are likely to forget is that they remain at last analysis the playthings of the irresistible forces of nature. The ground shakes a little, and thousands of human beings perish; a relatively small volume of poisonous gas spills over the rim of a crater upon a town, and the inhabitants are no more; the brute passions of men break forth and they rage like primitive savages. Underneath the artificialized life of man, so long as he remains, in the last analysis, an animal — which, so far as can be seen, will be for all his days on earth — flow the ungovernable currents of nature, in strength unimpaired. The presence of what look like senseless variations, sure to produce maladaptation to life-conditions, is proof of nothing except that either our unfavorable judgment of them is wrong or that selection is not yet in operation. If we wait till the prosperity declines and the "pinch" comes, this doubt will be resolved for us.

The primitive selection of the mores through the annihilation of their bearers is largely replaced, as civilization advances, by a mixed form, whereby together with the destruction of some of these bearers goes the subjection of others. This form has characterized all antecedent ages, and is potential at any time. Very many wars, whatever their ostensible causes, have been fought over the mores. The campaigns against piracy, from Pompey's time and before down to the conflicts of the Spanish against the Moros of the Philippines, are cases in point; so is the series of conflicts fought to destroy the slave trade and the slavery system. In cases of this kind, however, selection of the mores does not go to the full limit of annihilation; only so much of destruction is necessary as will reduce the strength of the adherents of certain mores, or will in such manner unfavorably alter the conditions under which they must pursue the struggle for existence as to lead to a modification of their offensive ways. Physical collision and killing as a mode of selection of the mores are a sort of last resort in these days, called for when other forms of attack on certain customs and habitudes are seen to have been of no avail. Naturally, then, they are likely to be put into operation as between groups rather than within any given well-amalgamated group.

We may now set aside the case of war to the death and take up the other broad alternative of the struggle between the bearers of different codes of mores — subjection. In so far as the outcome of this struggle — without which, as in nature, there could be no selection — was life or death, in so far was the ensuing selection found to be peremptory and thorough, resembling that in nature. But when the competition comes to have as its outcome on the one side mastery and on the other enslavement, manifestly the selection of the mores could not be so thoroughgoing and rigorous. Some of the mores of the conquered are sure to last on, for the intolerance of the victors must have its limits of vigilance if not of intent; in fact, certain of the mores of the subjugated might well be of a superior order, although in their *ensemble* they might have been proved inexpedient by the test forced upon them.

Such cases are for the most part instances of selection within a group. Conquest with enslavement creates the compounded group, and we begin here to talk about class-conflict rather than group-conflict. This is a developed form, for a type of societal organization must be somewhat advanced in order to render enslavement possible. In general, the case as respects the mores would run something like this. In the conflict preceding subjection, two groups of folkways have met and one group has been proved, to its adherents at least, superior. It is characteristic of the ethnocentric attitude of societies, as has been seen, to believe their own mores — their own prosperity-policy — to be best. But through this conquest the prosperity-policy of the prevailing group is vindicated. The conquerors have only satisfaction in their own system and contempt for the one whose adherents have fallen before them; compare the scorn of nomadic conquerors for subjugated industrial peoples. But it cannot be thought, on the other hand, that the conquered will regard their adverse fate as a condemnatory test of their code; this would assume a degree of reflective power not to be found under such conditions. Rather would they refer disaster to the aleatory element,[1] and so to some supernatural agency. But whatever they think, the situation impinges on them with power; and in the course of time the refractory will have been bent or broken, the less recalcitrant will have adopted, with or without compulsion,

[1] Sumner, " Folkways," § 6.

the ways of their masters, and the next generations will have grown up to accept the mores
popular or prevalent within their societal environment.   Ridicule, *i.e.*, an assault on vanity,
one of the most sensitive of human traits, has
always played an important rôle in selection
amongst the mores.   The tendency will be to
imitate the ways of the powerful [1] and so the
mores of the conquering group will tend unconsciously to prevail to the elimination of
conflicting codes.

The exact manner in which two sets of mores
belonging to two sections of a compounded
group will eliminate one another, or fuse, cannot be predicted; it depends upon the interplay
of many interests, which can never be the same
in any two cases.   But it is certain that the
process must have been easier under primitive
conditions, for the codes of neighboring peoples
could not have been very diverse.   To one who
reads ethnography one of the most striking
impressions is that of the essential likeness of
savage life and habitudes, not only within the
same region or continent, but all over the world.
It goes without saying that the mores of two
Congo tribes could more readily unite to form a

[1] Cf. pp. 114 ff. below.

composite than those of two advanced nations.
The result of advance in civilization is differentiation of groups and the development of distinctions between their habitudes.   Intolerance
of another religion, for example, is scarcely to
be found upon the primitive stage.   Even more
difficult would be assimilation as between the
codes of a savage and of a civilized group.

And so generalizations from cases of amalgamation cannot give us more than vague outlines of the process.   In any case it is the demonstration of the fact of selection rather than
its details that interests us in the present connection.   Spencer has a good deal to say of the
effect of homogeneity in strengthening the
social bond:[1] the best conditions exist where
the environment and race-character are of a
similar type, where there is a bond of blood, real
or fictitious, where similar religious ideas and
sentiments prevail.   "Joint exposure to uniform external actions, and joint reactions against
them, have from the beginning been the leading
causes of union among members of societies."
Gumplowicz,[2] doubtless with Austro-Hungarian
conditions in mind, insists upon the importance

[1] "Principles of Sociology," II, §§ 450, 451.
[2] "Der Rassenkampf," pp. 249 ff.

of a common language in securing amalgamation of ethnic elements. Language, he says, must first fall a sacrifice to the advancing group-sense — which he calls "syngenism" — for only by speech do men become *men* to one another. Religion, he says, is more stubborn,[1] but when it has once fallen, a great barrier is gone. Intermarriage, followed by community of blood, comes last, since the greatest conservatism and tendency to segregation reside here. This author adds to community of language, religion, and blood also community of culture and education and of material interests as providing the necessary conditions of social amalgamation.

The fact of selection in the mores succeeding the compounding of groups, whatever the method or order followed, is observable in all cases; and it is this fact which, for the time, engages our interest. After the groups have become amalgamated into one, changes appearing in the industrial organization, religion, language, etc., betray the fact that, without much conscious effort, some of the mores of the combining groups have been discarded and others generally adopted, thus creating a new code dif-

[1] The case of the Moriscos and Jews in Spain is illustrative of the persistence of the religious barrier.

ferent from either of the originals. If, now, there is a recompounding with smaller conquered elements, the prevailing code will gradually secure extension, showing contributions and substitutions from the mores of the conquered, but in the main remaining characteristic. There comes about a relatively bloodless selection in the folkways by which certain of them prevail against others. The whole process, it should be recalled, would be easier (and at the same time less distinctive in result) among the more primitive peoples, because of the lack of sharp and characteristic distinctions in their codes.

Assume, now, a society whose constituent elements are well unified, and which is at peace with neighboring groups. Since no society can be entirely homogeneous, the struggle goes on; but the internal issue as to the mores is concerned with less vital matters. It is a struggle between sub-groups which accept the broad code of their social union. The mores of the several sub-groups or classes, partly traditional, partly adaptive,[1] subdivide until they reduce to the family mores which form the basic categories. Each group has its code, which contains its peculiar elements and also

[1] Pp. 91 ff. and 251 ff. below.

others which it holds in common with other groups, and which form a basis for union, permanent or transitory, with those others. Certain relatively few of the mores remain common to the whole society; [1] these are by the general consensus regarded as vital, and members of the society who violate them are excluded from societal life (by execution, imprisonment, banishment, etc.). As the groups are subdivided in a society they show a more extended body of mores to which all in the local group or class are willing to conform (*e.g.*, those of sects); until

[1] "The rights of conscience, the equality of all men before the law, the separation of church and state, religious toleration, freedom of speech and of the press, popular education, are vital traditions of the American people. They are not brought in question; they form the stock of firm and universal convictions on which our national life is based; they are ingrained into the character of our people and you can assume, in any controversy, that an American will admit their truth. But they form the sum of traditions which we obtain as our birthright. They are never explicitly taught to us, but we assimilate them in our earliest childhood from all our surroundings, at the fireside, at school, from the press, on the highways and streets. We never hear them disputed and it is only when we observe how difficult it is for some foreign nations to learn them that we perceive that they are not implanted by nature in the human mind. They are a part and the most valuable part of our national inheritance, and the obligation of love, labor, and protection which we owe to the nation rests upon these benefits which we receive from it." Sumner, "The Challenge of Facts and Other Essays," pp. 353–354.

finally, in the family, the number of mores accepted by all members reaches the maximum. Down this scale the vital importance of the mores declines, so that, for example, many of the mores of the sect receive no general societal sanction, and a number of those regarded as of importance inside the family are viewed with complete indifference outside. This relation of the mores to the size of groups can be covered by the proposition that the number of accepted mores in a group varies inversely with its size.[1]

All these sub-groups, the more homogeneous and coherent as they are smaller, are engaged in a competitive struggle for self-realization, under the general condition of unification within the state and satisfaction with the national code. Considering the state as a single group, the great majority of its citizens will unite to support the general interest against the rela-

[1] "In general, the size and extent of groups stands in an inverse relation to the number of common syngenetic elements [mores] in them; so that the more populous the group the fewer are the syngenetic feelings common within it. The strongest, the most concentrated, so to speak, of the syngenetic sentiments, which rest upon the largest number of common syngenetic elements, are those which unite the smallest groups. The greater the groups become, the weaker are these sentiments, since they rest upon an ever smaller number of such syngenetic elements." Gumplowicz, "Der Rassenkampf," p. 249.

tively few recalcitrants who will not adapt themselves to the broad group-code. National public opinion will be against the latter and their ways; and in their cases selection will be decisive. Thus will the great majority of citizens in a civilized nation disapprove of polygamy or incest. But, whereas the body of the people will agree and coöperate in matters of broad national interest, under the general code to which nearly all subscribe, beyond this point divergence of interests and dissent are the rule. It is essential to the coherence of an amalgamated society that there shall be agreement upon what are conceived by powerful combinations of citizens to be the vital issues, or else there will be constant tension and civil strife; the indispensable condition is that the contentment over the provisions for what are conceived to be the paramount interests of all shall be profound enough to relegate the local interests of some to a position of minor importance. This allows of peaceful competition and rivalry and admits of toleration and compromise. Disagreement over this or that detail of the code does not become serious if all parties value the major provisions of the code. One of the chief utilities of otherwise inane and trite exhortations to patriotism and loyalty is that, by reiteration of these, the broad essentials of the common code and the need of supporting it are kept before people whose conflicting minor interests normally fill their horizons. They are made to realize that they and their interests are but part of a larger whole whose importance is of the greatest; and this and the tickling of their vanity help them to compose their local differences more readily.

Within the settled state the struggle between the exponents of divergent mores becomes, then, a different sort of competition from any we have yet viewed; and its prizes and penalties of a different and less vital order. Competition does not stop when violence is ruled out; it is only in the utopias of dreamers that it is eliminated. It can never cease, and this is well; for only with the realization of utopia would the race cease to have further need of the spur of necessity and of the stress which, through the ages, have made us move toward what we have become. This competition within the state is, in these times, in theory at least, no longer a struggle for existence, either of the individual or of the group; there are societal institutions and international codes whose busi-

ness it is to interfere in the struggle short of
that point.[1] That mere existence is somehow
assured is taken to be a fundamental fact; the
struggle is not for existence but for a certain
quality of existence; not for life, but for a
standard of living. Out of a form of competition thus mitigated no such clean-cut results
can be expected as are found in nature. This
struggle, in its separate cases, is that of the
individual or family, but it is carried on in
practice in some sort of wider association.
There is an unending series of identifications of
the individual or family interest with the interest
of this or that group, be it gild, union, sect,
party, class, or other. Let us take the class as
the most typical and least special of these
groupings of people with interests in common;
it is also an elemental one, developing as it does
out of the primordial master-slave differentiation. Gumplowicz [2] has developed this idea
of the conflict of classes in considerable detail.
The interest of competing classes is, at bottom,
a better standard of living — in the direct line
of their class-mores, but idealized somewhat.

[1] Cf. p. 307.

[2] "Der Rassenkampf"; also "Grundriss der Sociologie," and
"Sociologie und Politik."

There is a body of *wants,* common to all, and there is a coöperative effort to obtain satisfaction of these wants.

The right of claim to such satisfactions is the object of the struggle. We have a right to a thing when fellow-competitors who want it cannot or will not prevent us from gaining or keeping it. To avoid a beastlike struggle it is a condition that the rest hold off and let us have it. But a right has to be established in some way — has to be enforced by law or otherwise against fellow-competitors for satisfactions desired by many but not sufficient to supply all. Rights cannot exist where there are no fellow-competitors between whom claims to satisfaction are apportioned. On Crusoe's Island there was no occasion to define rights at all till Friday came. Rights are societal in origin and character. There is no such thing as a "natural" right; what are sometimes so-called are simply claims acquired so long ago and, among some sections of humanity, so strongly guaranteed and ingrained in the mores that they are no longer questioned.

The beginning of all rights or claims must have lain in force, or in the fear of its exercise. Even the rights of an infant are enforced by the might of the reproductive instinct projected at last analysis into mother-love. Those who could take and hold had the claim and the rest perforce submitted. Competitors would not of their own volition hold off from the appropriation of satisfactions; fear of some kind was the disciplinary factor, here as elsewhere. Rights were guaranteed and supported by appeals to that of which people were afraid: physical force, the power of authorities, of public opinion, of the ghosts and spirits that sanctioned the mores. Classes have always struggled to get one or more of these forms of power on their side, to enforce their "rights" against others. Until they secured one of them they had no rights at all; where slaves, for example, had none of these supports, they were without rights.

One of these forms of power has been singled out in history as the most obvious means of securing rights and the most evident sanction of them; this is political power. To-day talk about woman's rights, the rights of the laboring classes, etc., is mostly political talk; organizations to secure such rights are mainly political in nature, or soon become so. The privilege of making laws is the objective. There is a widespread conviction that laws once written on the statute books somehow acquire a power to alter the conditions of life as desired; and this in the face of archives teeming with legislation that was dead letter from the start, or was promptly repealed because of adverse effects. And yet the fact remains that the "governing class," if there is such, can exercise a certain selection in the mores.

A class is really definable only upon the basis of its mores; the code is the class. Terms like bourgeoisie denote a standard of behavior, a set of ideals, in short a standard of living, which is in the mores. Its code is the only distinctive thing about a class. But if a class gets political power, it can conserve and further realize its mores; and, since it is the powerful who are imitated, can very likely transmit its code to a wider social clientage.[1] The class, on the other hand, which has no political rights, has no chance to do this. Thus there is, in the class conflict for political power, a selection in the mores through their bearers. Probably the importance of this variety of selection is easily exaggerated, because it is an obvious form. It exists among primitive peoples in rather

[1] Cf. pp. 114 ff. below.

marked degree; for instance, in cases where the secret societies dominate the tribal life. It is a debatable question as to whether the imitation of the powerful is very thoroughgoing. This matter involves certain principles which will be met under the topic of transmission of the mores. But certainly all rights are not legal rights secured by control over legislation, nor does legislation alone confer rights.

With our present limited power of penetration into the operation of societal forces there seems to be, in this mode of societal selection, a lack of that precision, exactitude, and finality characteristic of a nature-process. Certain classes may be thought prosperous, powerful, and influential upon the life of society, but those discriminated against in the outcome are not eliminated from all influence, let alone from life. Even the slaves modified the language — one of the purest forms of the mores [1] — in the South. In Darwin's sexual selection the weaker male is not always destroyed, but simply to some degree excluded from passing on his qualities; and this sort of selection is regarded by Darwin as much less rigorous than natural selection. If that is so, then societal selection of

[1] Sumner, "Folkways," §§ 135–141.

the order before us is still less effective; for the unsuccessful class is to a still less degree prevented from transmitting its qualities.

What is true of classes in the broader sense is true in their degree of lesser societal groupings which are formed with the idea, sometimes almost unconsciously held, of emphasizing and propagating certain aspects of the mores; as, for example, sects and parties. They all strive for a position in the societal organization which shall enable them to pursue or enforce their particular prosperity-policies or pet social schemes. They gain recognition or vogue, it may be, or they are selected out almost before they have taken form. The fact of selection in all such cases seems patent enough; and it is reasonably plain, for this stage of the argument, that the struggle of classes, resulting as it does in the exaltation of this or that group to an influential position, results in some extension of the corresponding type of mores.

But if the seat of the mores is in the "masses," as Sumner demonstrates,[1] then we have not as yet sounded the depths of societal selection by considering the influence of this or that group

[1] Sumner, ' Folkways," §§ 52 ff.

which, for the time, has attained ascendancy. There may well be here an underlying, slow-working selection, which operates, when we come to understand it better, with no less precision and finality than its prototype in the organic field.   We are but at the beginning of the scientific study of human society and the way for a long distance ahead is beset with all manner of difficulties unknown to natural science.

In any case, the whole process is in the societal realm and seems quite divorced from the organic. Selection as the result of armed conflict was not so far from the nature-process; but the characteristic modern form of selection in the mores, though under it lies the nature-process ready to act *in extremis,* is on a different plane and of a different mode.

CHAPTER IV

IT has been important to get before us the general nature of the social conflicts without which there could be no societal selection, and the fact that selection takes place as a result of those conflicts. The nature of this selection was clear enough when it was a case of physical conflict followed by the extirpation of the conquered; but when the conflicting groups went on living together and amalgamating, finally forming a peaceful union, there was a change in the mode of selection which clearly distinguished it from natural selection and seemed to replace certitude and finality with vagueness and uncertainty. It is now desirable to clear up this uncertainty, if possible; and to do that it will be necessary to attack the matter from a new angle.

In organic evolution the sort of development that may take place is often largely a matter of what has already taken place; in a sense the

90

selection between variations is based upon the body of characters already assembled by the antecedent process. Certain lines of development are virtually shut off by reason of the direction along which the organism in question has evolved. The horse cannot now develop variations on the basis of five digits. If nature could be regarded as choosing and as giving reason for the choice, she might be figured as saying: "This variation must not be favored because it is not in the line of previous development." Subsequent evolution is largely conditioned by antecedent evolution. In societal evolution, as it appears within the group and under conditions that do not allow of violence, it sometimes seems as if the only selective force were of this order; or, to put it in the terminology here used, as if the only selective force to be exercised over variations in the mores lies in the antecedent mores. The great body of the society moves on in an inert way, judging everything on the basis of code, rejecting this and accepting that, for the most part irrationally, its mind made up for it by the traditional mores of the group. If this were strictly true, then societal evolution must have remained almost in abeyance since the formation of these

codes, and we might say that effective selection
stopped with the passing of the rude processes
presented in the last chapter. In fact, it is
borne in upon one who reflects much on these
matters that, in the life of society, the grand
feats of selection must have been accomplished
during the period of stern and bloody conflict;
that at that time the general course of future
development was, in a broad way, determined.
But, however this may be, it is also plain, upon
reflection, that the process of selection has not
ceased. The mores show adaptation [1] which
betrays selection still at work; and the codes
actually change, at least in less important par-
ticulars, before our eyes, Along with the strain
toward consistency goes another strain toward
adaptation of means to ends.[2] And it seems that
in some cases, at least, the means and ends are
visualized and that selection takes place on
rational grounds, and even in reaction against
the accepted code. In this country, for example, while there are certain things that must
not and cannot be done, owing to the mores [3] —

[1] Chaps. VIII, IX, and X, below.   [2] Sumner, "Folkways," § 5.

[3] Edith Wharton, in "The Fruit of the Tree," skilfully develops
a situation wherein a nurse, in order to spare a dying patient
useless suffering, helps her to die. Then the fact becomes known
and society wreaks its vengeance for this infringement of the

which public opinion will not yet tolerate, however rational they might seem (*e.g.*, cremation) — there are important movements which public opinion can be brought to support (*e.g.*, pure food laws). It is generally supposed that this latter result can be achieved through an appeal to reason; and the conviction seems popularly to be held that our social changes, alterations of code, and so on, are the results of the application of rational choice or selection. It is toward an unconfused and just appraisal of this aspect of societal selection that the present chapter is to aim.

A great many of the prescriptions and prohibitions in even the most primitive mores seem in retrospect to have been well advised for societal welfare. They seem to be rational.[1]

traditional code. A while ago a prominent judge gave cogent reasons for the conviction that, under certain circumstances, a man had a right to die; but such a proposition could not even be discussed, for it was contrary to the prevalent code.

[1] The following quotation, from McGee ("The Seri Indians," in *Bu. of Amer. Ethnol.*, XVII, pt. 1, p. 203), presents a peculiar illustration of this point. "Smaller rodents, especially the long-tail nocturnal squirrel, are excluded from the Seri menu by a rigidly observed tabu of undiscovered meaning. A general consequence of this tabu is readily observed on entering Seriland; there is a notable rarity of the serpents, the high-colored and swift efts, and the logy lizards and dull phrynosomas so abundant in neighboring deserts, as well as of song birds and their nests;

But this is true also in the case of animal ways, and yet animals cannot be said to exercise rational selection. That which conduces to success in the struggle for existence, and so is selected for perpetuation, turns out to be justifiable by reasoning subsequently applied. Circumcision doubtless began as a religious rite; later it was found to be hygienic in many cases; and presently the savage was accredited with the development of a rational practice. The mores of all civilized peoples have demanded an increasing security of life and property, and in the conflict of interests there has been no limit to the infliction of pain and death which the exponents of the prevailing code have been willing to put upon those of the aberrant type,

and this dearth is coupled with a still more notable abundance of the rodents, which have increased and multiplied throughout Seriland so abundantly that their burrows honeycomb hundreds of square miles of territory. A special consequence of the tabu is found in the fact that the myriad squirrel tunnels have rendered much of the territory impassable for horses and nearly so for pedestrians, and have thereby served to repel invaders and enable the jealous tribesmen to protect their principality against the hated alien. Seriland and the Seri are remarkable for illustrations of the interdependence between a primitive folk and their environment; but none of the relations are more striking than that exemplified by the timid nocturnal rodent, which, protected by a faith, has not only risen to the leading place in the local fauna, but has rewarded its protectors by protecting their territory for centuries."

to secure its elimination. Good reasons, accept-
able in the light of experience and knowledge,
could now be given for such action. But even
this does not mean that there was in that ruder
time any genuinely rational selection, deliberately applied. Even in the most modern of
cases it is most difficult to be precise as to the
entrance of this factor, for the rational and the
traditional or irrational are intertwined, and it
is very difficult to disentangle them; who, for
example, can analyze the attitude of the South
— or the North — toward the negro, and say
just what in it is rational and what irrational?
The "rational" is often no more than a subter-
fuge under cover of which the ancient "instinct"
or "second nature" gets its way — on the principle that the chief use of the human mind is to
find reasons, or subsequent justification, for
doing what its possessor wants to do. Thus the
form of societal selection which we see about us
in the modern age — and which must include
rational selection if it appears anywhere — is
in the last degree complex; it lacks all the clearness and precision of natural selection and of
societal selection in its ruder forms. Some say
that the process is all hit-or-miss; in fact, that
there is so little regularity about it that to try

to extend over the life of modern society such reasoning and theory as have set the facts of a less complex life in order is a vain and forlorn enterprise.

If selection is to be rational, it must be performed in the light of knowledge. The more extended and precise the knowledge, the more rational may be the selection. In order to choose rationally among variations arising in the mores, there is need of a perspective of the course of the mores. But this could not have existed in any significance among the uncivilized, for we find primitive man devoid of the imaginative power which would allow him to represent to himself future contingencies and plan to meet them, and not as yet in possession of a store of experience out of the past. If he thinks of the past, it is in the light of the code of the ancestors; if he questions the future, it is through magical means of divination. He is the "unfortunate child of the moment." [1] But the course of evolution led presently to the development of foresight [2] as the chief of the

[1] Fritsch, "Die Eingeborenen Süd-Afrikas," p. 418; Lippert, "Kulturgeschichte der Menschheit," I, 39.

[2] Lippert takes "Lebensfürsorge" as the basic mental trait in the evolution of civilization.

social virtues; upon the lower stages of societal evolution, that quality was of the utmost importance in the conflict of groups, and so became a criterion of selection. This meant the selection of the imaginative quality, the ability to foresee and plan. But this amounts, in time, to the development of the power to see a perspective of individual, and at length, of group-development; and to the rise of a consequent consciousness of the course of such development. But if a course can be consciously visualized, it can be criticized, and the future can be laid out in the light of possible corrections of a faulty procedure. With the development of writing, a record of the process can be kept. There can be formed a body of ideas or knowledge about the conditions of life which can be held true, subject to correction. In other words, laws can be derived from the instances preserved. The *ensemble* of these is science; and, in its most perfect form, science (as "trained and organized common sense") affords a basis for rational prediction, a thing very different from divination. It is by science that one gets as far as possible outside of the mores of his time and local group, and wins a vantage-ground from which to pass rational judgment on the

H

code. Knowledge about the laws of nature and of societal life is that which enables us to adapt ourselves more successfully to these laws and so to avoid the pain and disappointment that comes from the attempt to live out of conformity with the inevitable. This is the only sort of theory upon which a belief in rational selection could be based.

The process outlined is precarious and full of error; for where the human mind, with all its inefficiencies and limitations, comes in, there are sure to be wrong valuations due to imperfect knowledge and to bias of various kinds. The most emancipated of men live largely in the mores and cannot free themselves from their domination. And it is well that they cannot, for, as Sumner[1] remarks: "if we had to form judgments as to all . . . cases before we could act in them, and were forced always to act rationally, the burden would be unendurable. Beneficent use and wont save us this trouble." The sum of real knowledge, especially of that outside of the field of use of natural forces, reveals itself repeatedly as being exceedingly small; yet our folk-wisdom recognizes that knowledge is power. Where knowledge has

[1] "Folkways," § 68.

been checked in its course, power in the conflict of life has not increased as it has where knowledge has been on the constant increase. Compare the present status of China with that of the western nations, and then her status centuries ago, when she led the world in knowledge, with that of the ancient Germans and Gauls. Compare the Japan of 1910 with the Japan of 1810.

Within, as well as between, groups or nations it is proverbial that knowledge comes to the top; if a father wishes his son to occupy an influential position in life, he educates him. Every one knows that the great material advance of the last decades has come out of science; the son must know science. Likewise must he be acquainted with the so-called cultural studies; for a boor must be a veritable genius to be tolerated by those whose interest means success in various ways. Much that is taught is untrue or trivial, but there is a residuum of generally accepted truth which is indispensable, and which can be taught to all who are capable of learning. Its effectiveness varies with its degree of positiveness and verifiability. It serves as a touchstone on which to test some at least of the mores. There are certain mores which an educated public opinion will not tolerate, for example, expectoration in public places; science has shown that they cannot possibly be good for welfare. These must pass with the advance of science and education; in fact, they are selected out before the body of the population know that they are harmful; for those who know are powerful out of· all proportion to their numbers, and are able to suppress this or that habitude by the use of the regulative system. The selection exercised by positive knowledge in the folkways may be very erratic and ill-advised in the light of more knowledge; but we are not passing judgment on it yet. The idea at the present point is to demonstrate the fact that the progress of conscious mental activity may lead to rational selection in the mores where it could never have existed before.

It is not proposed, however, to concede any extensive direct influence to rational selection over the mores as a whole; it is indispensable, rather, in order to represent things as they are, to throw into some prominence the essentially irrational and traditional nature of the mores. For one who observes the life of the society he lives in without more bias than flesh is heir to, a few words by way of illustration, and a reference to the encyclopædic chapters of "Folkways" are all-sufficient. It has been said that most of us live for the most part in the mores, acting unconsciously after their prescriptions; and that all of us live to a very large degree in them; that is, that we are all mainly creatures of use and wont. Few of us have a positive, well-thought-out plan of life, whether it be right or wrong. We marvel at the self-directed life of a Goethe, and assign it forthwith to genius. We meet emergencies one or two at a time, acting in them for the most part without much perspective beyond the case in hand. There are too many exigencies for us to give much thought to each, and so we go on by rule of thumb, that is, in the mores. So long as we slide over vicissitudes, or slip past, that satisfies us. If our mores effect for us a painless passage through life, we have no idea of criticizing or altering them; we let well enough alone. Every man, no matter how enlightened, has his point where it does not seem worth while to cerebrate any further, and he "takes his chance." A Newton could concentrate "a little longer" than the rest of his fellows; the savage leaves off somewhat earlier in the process, just as the child does;

but we are all only human beings with minds that tire and attention that flags. Then we fall back unconsciously on the ways that have come out of the experience of the race. Consequently, while the advance of knowledge has exercised selective power over the folkways, and will probably wield an ever increasing influence upon them, yet the bulk of selection must still be of an automatic and unconscious order; since the great majority of men live almost entirely in the mores, and the great majority of the actions of the most enlightened men are likewise determined by them, the rational selection, of which we have spoken, does not go far in explanation.

What is true of the individual is even more marked of the group. Seldom does a group deliberately set out to apply rational criticism to its mores. Our sex taboo is largely irrational, inherited as it is through religion from dogmatic propositions of people who had no authority for us. A Hindu lends his wife to get good offspring. This may be rational on the face of it, but it is also shocking, being against our mores, and most of us would not even tolerate the discussion of its rationality. There is probably a good reason for our repugnance, but we do not stop to reason. It seems rational in every way that a widower should marry his deceased wife's sister, especially if he has young children. Such a second wife would naturally be least likely to turn out a "viper to the first brood"; and there is, of course, no more question of incest than there was in the first marriage. But the vicissitudes of the Deceased Wife's Sister Bill in England show how difficult it was to make reason heard, even in so clear a case where the mores were wide open to criticism. In short, as Sumner says,[1] the mores can make anything right and true and exempt it from criticism.

The primitive touchstone for the mores was the sensation of pleasure and pain, or, to put it more generally, the degree of satisfaction experienced. It was all empirical and without perspective beyond the case in hand. This is characteristic of the action of men under the mores as distinguished from action based upon scientific knowledge. It is present dissatis- faction rather than the anticipation of a better satisfaction — a negative rather than a positive incitement — that leads to change, involving selection, in the mores. And the dissatis-

[1] "Folkways," §§ 31, 32, 80, chap. XV.

faction or need must be felt strongly if it is to provide an incentive capable of overcoming all the inertia and superstition that block the way to alteration in the code.

This dissatisfaction led, at the outset, to "blundering efforts" to satisfy the need; but the case is not so much different, upon higher stages of culture, when it is a matter of old and settled folkways. The stronghold of the mores is in the "masses," [1] and their efforts to escape from their dissatisfactions have always been clumsy and often aimless. Sometimes they do not even realize that they are unfortunate and miserable — especially if they are isolated from the rest of the world, as the Russian peasants were, and so have no standard of comparison — and when they are roused by their pain they plunge about with no settled plan and no rational aim. A skilful demagogue can lead them in almost any direction, by playing on their inflamed feelings and then holding out hopes which are seldom rationally criticized. If the dissatisfaction is keen enough there comes about, often with bloodshed, a revolution and a new type of mores, which, whether it yields satisfaction or not, amounts to an alteration of

[1] Sumner, "Folkways," §§ 47 ff.

the incidence of discontent, so that at length the new order, by contrast with the unsettled period of revolution, seems in some degree satisfactory. In any case there is change, and the monotony is relieved. There is a selection of some kind in the mores.

Such upheavals on the grand scale, as in the French Revolution, are repeated on the small scale from time to time in any state. Some one has said that every presidential election in the United States is a bloodless revolution; some of them have barely stopped at that point. This brings before us the struggle of two composite sub-groups not necessarily conterminous with social classes: political parties. In a general way political parties grow up about the nucleus of a common interest that is felt rather than reasoned out. Party affiliations are largely in the mores and are traditional. They are the expression of an interest or interests held in common by smaller groups which are willing to unite for their realization. In other respects such uniting groups may have very different codes; "politics makes strange bedfellows." However, political parties are likely to comprise those who have a number of interests in common, or think they have, and so are often co-extensive with classes or even geographical sections.

These diverging interests of parties are settled for varying periods by elections, and the methods of these reveal the essentially irrational and even primitive nature of the selection of a new prosperity-policy. In general, the party in power will stay in power so long as it can "give the people prosperity," or persuade them that it is doing that — so long as they are contented. The election plea at its weakest is to let well enough alone. There is never any candid reference of prosperity to non-political factors, as, for example, the rich natural resources of the country, its underpopulation, or the like. It is always the pet policy, protection or free-trade, war or peace, that has, singly and alone, and as the result of wise and rational procedure, produced well-being. History and common sense are perverted, deliberately or under bias, to demonstrate this. There is a great deal of "pathos" in the party system, which betrays its fundamental relation to the mores. Pathos, in the original Greek sense of the word, says Sumner,[1] "is the glamour of sentiment which grows up around the pet notion of an age and

[1] "Folkways," §§ 178, 179.

people, and which protects it from criticism.
. . . Whenever pathos is in play the subject
is privileged. It is regarded with a kind of
affection, and is protected from severe examination. It is made holy or sacred. The thing is
cherished with such a preëstablished preference
and faith that it is thought wrong to verify it.
Pathos, therefore, is unfavorable to truth."

On some such body of sentiment in the mores
rests the "solid" strength of a political party.
But to hold its dominance it must, like the
Indian medicine-man, prevent misfortune. No
matter what the source of "hard times," their
appearance is ominous for the party in power.
It is ineffectual to demonstrate that "we didn't
do it," even if that were possible in the face of
the accusations of the parties out of power; the
answer would be: "Why didn't you stop it?"
and there would be no query tolerated as to
whether it was in human power to stop it. This
condition of things is often jocularly expressed
by referring the cause of floods or droughts or
earthquakes to the inefficiency of the ruling
party.

A state of popular discontent, then, leads
politicians to expect an overturn, for the people,
in this country at least, having in their mores a

great faith in government activity, think to better their condition by voting out the party in power. An appeal preceding election to the mores of the several classes and sections, in the platform of parties and in the utterances of party men, is a characteristic part of procedure. There is some show of appeal to reason, for that flatters audiences, but the main line of attack is along the *argumentum ad hominem*, which is an appeal through the emotions to the mores. The "straddle" is characteristic. The fact that the *post hoc ergo propter hoc* argument is the stock in trade of the parties in assault and defense betrays the essentially non-intellectual character of the whole proceeding. Each candidate is endowed by his supporters with the standard virtues of the mores ("honesty," etc.), and often the attempt is made to "blacken the character" of the opposition candidate by reference to delinquencies of one kind or another. Our man is the "man-as-he-should-be" [1]; the other candidate is not; when we talk about the former to people of a certain code we endow him with the local virtues which we deny to his antagonist — the latter takes wine, which our man never does. But in certain other quarters

[1] Cf. Sumner, "Folkways," §§ 206, 207.

we assert that our man is a jolly good fellow, whereas the opposition candidate is a dull prig, a man of ice. Thus do we strive to enlist prejudice by appealing to the mores, in order to carry an election, that is, to bring about a selection in the mores. And yet this scheme of procedure is, almost by common consent, conceived to be the fairest and most enlightened way of securing such selection. Every one has his vote, we say, under the modern democratic system, which is itself the product of selection through tremendous social upheavals. There is, indeed, a "pathos" of democracy, but in view of its spread over the civilized world, we are almost forbidden to believe that it is not a better adapted form than those which preceded it. And certainly no better case for rational selection could be made out for the antecedent systems.

If we seek to focus the light thrown on selection in the mores by what has been said, we find that dissatisfaction leads to an effort to secure change, and that the effort is directed in a more or less blind way toward the better realization of local and personal interests. The latter type of interest is almost if not always at the bottom, but it can appear legitimately only as a group or class (societal) interest. This is the class conflict, the clumsy straining after a better, or a less bad standard of living, somewhat idealized out of the class mores; *e.g.*, a "full dinner pail" for some, more "protection for national industry" for others, and so on. There is always the promise to "do something" for the farmer, the laborer, or for some other class; that is, to ease up the competition in which he is engaged with his fellow-men, and to elevate his standard of living. If this is not societal selection in his favor, it is hard to give it a name. But these beneficiaries seldom approach the "peaceful revolution" in a rational state of mind; at best their theory is the vicious one that the state can do anything, and they propose to tell it what to do, at least for themselves.

Here would be, perhaps, a proper place to consider the demagogue or political boss. Says Woodrow Wilson [1] of the latter:

"The man we call boss is the agent of those who wish to control politics in their own interest. I have known some of these gentlemen personally, and I know exactly how they work. They haven't any politics at all. That

[1] Address to Princeton students, in the Associated Press report for Sept. 26, 1912.

is the point, and there is no difference between a democratic boss and a republican boss, because neither of them
is working for his party.   They are both working for their
clients."

And it is needless to say, to American readers
at least, that the methods of the boss are as far
as possible from broad and rational policy.   The
function of the demagogue in a democracy is
outlined by Lecky [1]:

"Every one," he says, "who will look facts honestly
in the face can convince himself that the public opinion of
a nation is something quite different from the votes that
can be extracted from all the individuals who compose it.
There are multitudes in every nation who contribute
nothing to its public opinion; who never give a serious
thought to public affairs, who have no spontaneous wish
to take any part in them; who, if they are induced to
do so, will act under the complete direction of individuals
or organisations of another class. . . .   And in a pure
democracy the art of winning and accumulating these
votes will become one of the chief parts of practical
politics.

". . . The demagogue will try to persuade the voter
that by following a certain line of policy every member
of his class will obtain some advantage.   He will en-
courage all his utopias.   He will hold out hopes that
by breaking contracts, or shifting taxation and the
power of taxing, or enlarging the paternal functions of

[1] "Democracy and Liberty," I, 22–23.

government, something of the property of one class may
be transferred to another. He will also appeal persist-
ently, and often successfully, to class jealousies and
antipathies. All the divisions which naturally grow out
of class lines and the relations between employer and
employed will be studiously inflamed. Envy, covetousness,
prejudice, will become great forces in political propagand-
ism. Every real grievance will be aggravated. Every re-
dressed grievance will be revived; every imaginary grievance will be encouraged. If the poorest, most numerous,
and most ignorant class can be persuaded to hate the
smaller class, and to vote solely for the purpose of
injuring them, the party manager will have achieved his
end. To set the many against the few becomes the chief
object of the electioneering agent."

Professor Sumner used to dilate upon the
absurdity of setting up a question upon which
experts were divided (e.g., the tariff, the currency system) for several millions of at most
partially informed people to vote on; [1] that

[1] "Propositions as to public policy," says Holmes (in Vegelahn
vs. Guntner, 167 Mass., 92), "rarely are unanimously accepted,
and still more rarely, if ever, are capable of unanswerable proof.
They require a special training to enable any one even to form
an intelligent opinion about them. In the early stages of law,
at least, they generally are acted on rather as inarticulate instincts
than as definite ideas for which a rational defence is ready."

Sumner writes of "legislation by clamor" as follows: ("The
Challenge of Facts and Other Essays," pp. 186–187): "On the
one hand, the highest wisdom of those who want anything now
is to practise terrorism, to make themselves as disagreeable as

they do not vote on the merits of the case, the sectional division of the parties (as in 1896) would indicate. They vote for what promises to come nearest to satisfying their interests. From the standpoint of one who is looking to see rational selection applied in the development of intelligent policies, this is indeed absurd. And it becomes the more ridiculous when the

possible, so that it shall be necessary to conciliate them, and those who appeal to reason find themselves disregarded. On the other hand, the public men seek peace and quiet by sacrificing any one who cannot or does not know enough to make a great clamor in order to appease a clamorous faction. It is thought to be the triumph of practical statesmanship to give the clamorers something which will quiet them, and a new and special kind of legislative finesse has been developed, *viz.*, to devise projects which shall seem to the clamorous petitioners to meet their demands, yet shall not really do it.

"The most important case of legislation of this kind which has been passed in this country is the Bland Silver Bill. It contains no rational plan for accomplishing any purpose whatever. It never had any purpose which could be stated intelligibly. It does not introduce the double standard, does not help debtors, and if it favors silver miners at all, does so in an insignificant degree. It satisfies the vanity of a few public men, quiets the clamor of a very noisy faction who did not know what they wanted and do not know whether they have got it or not, complicates the monetary system of the country, and contains possibilities of great mischief or great loss. It was passed as a patched-up compromise under the most rhythmical and best sustained clamor ever brought to bear on a public question. Those who raised the clamor went off content because they thought that they had obtained *something*, and they now resist the repeal of the law because they would feel that they had lost *something*."

very mechanism of the election is, as it were, devised to thwart intelligent voting, and to throw the management into the hands of specially interested parties.[1]

If assent is given to the preceding views, it cannot be asserted that rational selection prevails in any comprehensive degree throughout the evolution of even the most civilized societies; the masses are far from being saturated with knowledge, even in the most advanced countries. And in the masses, as Sumner demonstrates,[2] is the seat of the mores. But "the leading classes, no matter by what standard they are selected, can lead by example, which

[1] "It may be said without reserve that the sound reform of the ballot is a condition precedent to any real and lasting improvement in the management of political affairs throughout all the Union. It is absurd to ask the voter to pass on principles or policies, however good in themselves, when he is so handicapped by a clumsy ballot and frequent elections that it is physically impossible for him definitely to record his will. It is useless to urge him to vote only for good men and good measures when the conditions in which his vote is cast reduce him to a blind guess as to the net result of the vote. And we especially urge on those who are sincerely convinced that more 'direct' share for the voters in the Government will accomplish needed changes that no really direct or effective share in government can be attained until the short ballot is secured. You cannot get rid of the professionals when you leave the mechanism of politics so intricate and confused that only professionals can run it." *New York Times*, May 17, 1914.   [2] "Folkways," §§ 52 ff.; § 99 (quoted).

always affects ritual. An aristocracy acts in this way. It suggests standards of elegance, refinement, and nobility, and the usages of good manners, from generation to generation, are such as have spread from the aristocracy to other classes." Possibly there is here some opportunity for the exercise of rational selection by the leading classes, which will be imitated by the rest. But, Sumner goes on, "such influences are unspoken, unconscious, unintentional. If we admit that it is possible and right for some to undertake to mold the mores of others, of set purpose, we see that the limits within which any such effort can succeed are very narrow, and the methods by which it can operate are strictly defined. The favorite methods of our time are legislation and preaching. These methods fail because they do not affect ritual, and because they always aim at great results in a short time. Above all, we can judge of the amount of serious attention which is due to plans for 'reorganizing society,' to get rid of alleged errors and inconveniences in it. We might as well plan to reorganize our globe by redistributing the elements in it."

The truth of these remarks is evident. Schemes like the one projected a few years ago, to win the world to Christianity (Protestant Christianity) within a generation are as fanciful as any perpetual motion project. But if there are ways in which even modest results can be secured, these should not be passed over. Following the suggestion in the passage just quoted, it would seem that the most promising place in the society to apply rational direction to societal life is among the "leading classes." This brings us around again, but along a somewhat different avenue of approach, to the question discussed above,[1] as to the influence of a dominant class upon the mores of other classes in a society. Only now the query is a little more precise, for we are now confining our attention to rational selection. One point here is quite clear: the dominating class has a better chance than the rest to get knowledge upon which to base a rational selection. It is better educated.

There is not much sense, though, in the term "leading classes," unless that means that such classes have control over the policy of the society, that is, influence over its regulative system. But if we have the educated in power, then do we not have the class best fitted to

[1] Pp. 73 ff., 86–88 above.

exercise rational selection in a position to exercise it? This cannot well be denied; but then we are thrown back on the general case and have to recall what we have found out about the power of a dominant class to effect selection. The matter of its competency, to do that, scarcely enters into the question of its power. However, if it has even a little more power than the rest — and we have seen that it has — it is comforting to feel that it is better equipped for the function.

This is in some respects, and despite very evident and discouraging examples to the contrary, a rather hopeful view. Let us consider it in perspective and see what there has been in it. What has been the influence of the aristocracy of the educated, and what are its limits? The chief and medicine-man at the outset, even when they were not one and the same person, represented the knowledge of the group; the councils were the experienced members of the society; and so the destiny of the latter was intrusted, so far as it depended on reason at all, to those who rose to the top. These had behind them, then, the political power — it was the prize of the intra-group (class) struggle, to which, by reason of some superiority, they had attained. This they could use to control the rest; and provided they did not too directly oppose the mores residing in the "masses," they could make this and that change toward what they thought expedient. They were a minority generally, but well-knit and organized, and so were able to impose their will, within limits set by the mores, upon the majority. If they went beyond these limits, they were overthrown; but they gradually learned not to do that. They had certain mores of their own and certain ideas about the mores expedient to the societal welfare, and they were able to enforce a certain rational selection upon the whole group. If they acted tactfully they could crowd out certain of the mores, sometimes by almost imperceptible degrees, and could foster others in the same way. The fact that they were, in their position, objects of imitation,[1] assisted them somewhat in this direction.

This is true in any age, and can be seen to-day as of old. The mores of subjection to authority have strengthened the process. Respect for those in authority over us is inculcated from early years; and also we render unto Cæsar

---

[1] P. 217 below.

the things that are Cæsar's because it is easier to do that than to put up an unorganized resistance against an organized power. Many of our laws rest upon a clearly rational basis and are regarded as extremely tyrannical by our newer fellow-citizens; but they are obeyed because they are law, and after a while habit comes in, and the tyranny is felt no more. We learn to adapt ourselves to a steady and not too severe stress. Probably every one of our laws would have seemed intolerable at some preceding rude age, and many laws, the German for instance, would so seem to members of another nation. At first come irritation and evasion; but if enforced for a time, and not too violently in conflict with the mores, the regulations are accepted, and for the most part unintelligently, by the bulk of the population, and taken as a matter of course, as a condition of societal life than which no other is known. Where, then, the educated hold a position that enables them to "sway public opinion," they can exercise a certain amount of rational selection over the mores. If their ideas can no longer secure the effective sanction of religion, they can at least secure that of the state. By the exercise of such stress there may then occur for the body of a society, supposing the measures of the leaders to have been rationally conceived, a certain indirect rational societal selection. This line of argument might be pursued still farther. The lower classes of to-day certainly rank the leading classes of a past age in point of knowledge, at least in some departments of life; will they not, therefore, in the course of time, secure such share of the mental outfit as will enable them to modify their own codes for themselves in accord with reason?

But all this is possibility rather than actuality; the day of general intellectual emancipation is yet afar off. Even for the best educated the power of rational selection, as directed toward the destiny of the society of which they are a part, is as yet an uncertain and unreliable one in most departments of societal life. It is a source of never ending astonishment and disheartenment to observe the ignorance, prejudice, and even superstition displayed in politics by people who are well-informed and rational of judgment in other lines. Yet it is perfectly natural, after all; the human mind has its ever-present limits and cannot cover much ground thoroughly. It is difficult to look at the large interests of society in a large way, on their merits, for the lesser interests of person, group, constituents, etc., fill the foreground and hinder perspective. Furthermore, we all live so fully in the mores of our limited group that it is next to impossible to set ourselves in others' places and correctly and dispassionately appraise their interests. It is asking too much of human beings to expect one group to safeguard the diverging interests of another. Each group must try to get power to realize its own. It is the exceptional statesman who can rise above party and local interests; the perfectly respectable representative of a constituency with peculiar interests will insist upon these in the face even of party leaders, as witness the attitude of the Louisiana senators when the sugar tariff has come under discussion. Whatever lofty and rare statesmanship may be, politics is the play of local interests; for the realization of the interests of the society at large something more powerful and elemental than the ratiocination of individuals or parties must be in operation.[1]   Tak-

---

[1] If any one is enthusiastic over the influence of the "intellectuals," let him reflect that they could not introduce even a metric system, however strong their appeal to pure reason. And yet, in their automatic evolution, numeral systems show most satisfactory examples of variation and selection. The development

ing the field of societal life as a whole, one is reluctantly forced to admit that, in default of much power of rationalization, we are still thrown back overwhelmingly upon the mores, with all their apparent unreason and caprice.

This cannot in candor be denied. Rational selection can come about only to a slight degree as the result of the dominance of any single class or sub-group, let alone individual. Societal selection is too massive a phenomenon to take its origin elsewhere than in masses of men; it is due to the largely unconscious contortions of the whole social body. It is limited by conditions in the physical and societal environment, the power of which the would-be innovators are forced to acknowledge when at last they attain the coveted control. Note how the opposition party, the Jeffersonians,[1] for instance, is obliged to continue the policies against which it has been inveighing.

Says the sagacious printer-statesman:[2] "Those who govern, having much business on their hands, do not generally like to take the trouble

of the symbol zero carried a whole system with it. Cf. Smith and Karpinski, "The Hindu-Arabic Numerals," especially chap. IV.

[1] Sumner, "The Challenge of Facts and Other Essays," p. 329.

[2] Franklin, B., "Autobiography" (ed. Aiton), p. 228.

of considering and carrying into execution new projects. The best public measures are therefore seldom *adopted from previous wisdom, but forced by the occasion.*" It is a commonplace that the "ins" are conservative, just as the "outs" are radical. The former have gotten and are maintaining their position, and are serving their interests best, under the *status quo.* This is one of the main reasons, rising out of the depths of human nature, why adaptive changes are brought about, at the hands of the dominant class, less frequently than one would expect at first sight.

Let us cease considering, therefore, the constituent human elements of a society, and attack the issue on a broader arena. Let us consider the society as a whole, living a life analogous to that of an indivisible organism, and examine the functioning of the whole rather than its constituent parts.

Here we must notice a point of view which I can best introduce, perhaps, by a quotation.[1]

A professor of economics, we are told, "has expressed great astonishment at the fact that the currency bill was passed in so excellent a form by a congress in which 'not

[1] *Collier's Weekly* (quoted in *New Haven Journal-Courier,* January 20, 1914).

10 men understood the subject.' This is the expert's contempt for the intelligence of the average man. The specialist is always astonished at the ' luck ' which ordinary people have in making up their minds correctly.

" And yet this ' luck ' is democracy's only hope. If the processes of government are too complex and subtle for the average mind, then such a thing as a' people successfully safe-governed is out of the question. Lawyers often condemn the jury system because the jurors are not ' specialists ' in evidence; but a careful study of jury verdicts will show that their decisions are as good as or better on the average than those of judges on questions of fact where such questions are submitted to the court. Congressmen are, to speak within bounds, rather above the average in intelligence, and greatly superior to jurymen; but, if, as [the professor] suggests, they were ignorant of the questions involved in the currency bill, they were about in the same position as jurymen seeking the truth honestly and ignorant of the law. The one receives the instructions of the court; the other is favored with the expert instructions of [the professor] and others. The jury takes it upon itself to ignore the instructions once in a while, and the congress refused to accept the opinions of the professors when told by them."

It is implied in this quotation, which would command considerable assent, that special knowledge does not count for much, and that the "common man" is a repository of a sort of wisdom by intuition which enables him to decide correctly on national policy — that is, to perform an act of societal selection. If we discard the demagogic character of such assertions, and try to see what of truth there is in them, it amounts to this: the legislators represent the interests of different sections of the society; they do not, as we have seen, as a rule, try to visualize the national interests concerned, but vote as representatives of their constituencies. They are the organs of transmission, therefore, of public opinion. We are brought to the proposition, then, that public opinion is right where special knowledge may easily go astray.

One way of justifying this view about the correctness of public opinion is conveyed by the term "luck" as used in the quotation. Luck is a term, like chance, used to cover that which is, for the time, unreckonable and unpredictable. But attempts are made, in insurance enterprises, for example, to reduce the variable called chance or luck to a constant. This is done by applying the laws of chance to multitudes of instances, and it is thereby seen that pluses and minuses cancel out, leaving a residue upon which such dependence can be placed that business can be done with security. Now some people think that a similar residue of dependable truth is gained from the cancellation of opposite or partially conflicting judgments of multitudes of people, irrespective, apparently, of the rationality of such judgments. This is merely a matter of faith, with the chances against it. It has been proved wrong plenty of times in the outcome, based, as it has often been, on ignorance and superstition. But this does not say that it is always wrong when it opposes the conclusions of the specialist, nor yet that it is equally unreliable upon all varieties of issues. No outside observer, however learned, can sense interests as those can who feel their stress directly; and upon some interests men have a sounder judgment than upon others. Hence the conflict of interests, especially in the region of self-maintenance, where, as we are to see, they are concrete, objective, and material, may result in a wholesome selection of measures that no single specialist could have the breadth of outlook even to visualize in their multiplicity and complexity.

This somewhat anticipates what is next to come. We are now considering a society as a whole and have given up any idea of consistent rational selection in its mores at the hand of any dominant sub-group. Granted, however, that there is no class or other limited group through which radical rational selection is realized; is there not some point of entrance into the societal system itself, some part of the code, common to all constituent groups in a society, which admits this factor? I believe there is, and that its determination opens the only vital point of attack for rational selection. Rational selection in the mores, as applied constantly, directly, and generally to all societal forms, cannot be demonstrated; but I think it will be seen that such selection is natural and effective in certain of the societal forms and then, strongly rooted in the very seats of society's life, is enabled to extend its effectiveness, though indirectly, throughout the whole societal structure. In order to render this view, toward which we have been working, as clear and definite as possible, it will be necessary to approach the issue from a selected point outside, and somewhat by way of a detour.

CHAPTER V

To act with reason, or science, or common sense, that which is most needed is verification. Verification, unmistakable and also repeated, will prove anything. Neither logic nor prejudice can stand before it long; it is after the facts of repeated experience that science comes limping. Such verification can be attained in the field of the natural sciences through experimentation, where the environment can be controlled and artificialized and the object of experimentation manipulated. This, we have seen, cannot be done in the domain of social science; it is forbidden that man should experiment on man. What goes by the name of social experimentation is generally vague and inconclusive in its results, and error must needs be rife. This is one of the reasons why the social scientist looks with keen envy upon the procedure, yielding tangible results, of his colleague in the natural sciences. The best the social scientist can do is the worst the natural scientist has to do — to wait on nature and history to perform quasi-experiments for him; and he seizes upon one of these, like that exhibited in the case of the Pitcairn Island society, with pathetic eagerness. But one can profit only partially even by recorded previous experiences, in view of the loss of detail, the omission of that which is later seen to be of commanding importance, etc.; and then nature next to never repeats societal phenomena in the same terms. She even seems to act furtively, as it were, so that men do not recognize the scientific significance of what she has been doing till too late. The social scientist must get the best tests he can, and put up with their insufficiencies. We must know the best places in which to look for the occurrence or repetition of social phenomena; and in a hard matter like this one of societal selection, the favorable field of observation must be sedulously sought.

Where, then, in the societal system, can verification be found? Where shall the student look to discover and verify the activity of rational selection based upon positive knowledge? Where in the societal field are inadequate adaptations to the environment, as exhibited in unfit mores, most irresistibly demonstrated to be

K

inadequate and unfit? It is a fact of observation that dissatisfaction due to the non-realization of human interests may occur in an equal degree anywhere in the societal field. But, whereas in some parts of that field it seems almost impossible to reason out conclusively the causes of dissatisfaction and then remove them, in the case of others there is a general persuasion that the nexus between effect and cause can be passed and satisfaction attained — and that through rational processes unaided. It is a bold man who believes that through rational means the race can be bred so as to be free of the physically and mentally weak; but every one expects as a matter of course that fifty years will see the solution of many a present and arising difficulty in the line of material progress. Nobody in 1890 could reasonably expect the abolition of prostitution within any definite period; but there were those who regarded aërial navigation as merely a question of time. Men look to and depend upon rational procedure all the time in some parts of the societal field; they take up the war against smallpox and typhoid with confidence in their ultimate success. They realize that there are probably unknown physical elements, and are not forced to regard that man as a wizard who discloses them; they expect more surprises like the discovery of radium and the development of wireless telegraphy. But they have no hope that their religious mores will be adjusted as the result of some scientific discovery of a positive character; or that sociologists of a certain type may be able to set the relations of the sexes upon a sure basis. They do not hope for inventions in religion and marriage which can be tested with certainty and introduced with general or intelligent approval. Such results will come, they say, in the millennium — that is, when the conditions of life as now lived shall have been altered as the result of an unpredictable or even supernaturally directed development. Examples of this sort crowd to the page. It is not hard to demonstrate to an ignorant person in this country that he should learn to read and write; he can see that by living in this society. Similarly for his interest is it that he shall use the English language. Tests lie all about him, and are immediate and decisive. But try to persuade him by abstract argument to give up the vendetta, to renounce anarchistic leanings, or to change his religion, and you fail. There are no immediate and decisive tests at hand. You cannot demonstrate that interest will be subserved by change; you cannot even secure visualization of evil consequences. Even illness due to filth, where such visualization is becoming more practicable, can be referred unverifiably to too many different causes, as, for instance, the evil eye.

If we seek to generalize from the multitudes of instances on the order of those just given, the only conclusion at which we can arrive is, that the more nearly custom (the folkways) represents direct reaction on environment in the actual struggle for material aids to existence, the more rational a test does it undergo; and, conversely, the more derived the societal forms the more clearly do they fall under the tests of tradition rather than reason. The nearer the mores come to the struggle for existence, the more nearly they concern self-maintenance, the more vivid is the demonstration of their expediency or inexpediency.

But we must turn aside for the moment to get before us some perspective of the activities of a human society that we may observe which of them represent the most direct reactions upon environment. De Greef [1] has made a classi-

[1] "Principes de Sociologie," p. 214.

fication of societal phenomena which is as useful as any for our present purpose. He divides them into the economic, the genesic (of marriage and the family), the artistic, those relative to beliefs, the moral, the juridic, and the political. It is plain enough, without involving ourselves in this author's balancings and cross-classifications, that the first of these categories contains the bulk of societal phenomena that submit to ready and conclusive verification. Sumner classified the activities of society in a less meticulous manner into societal self-maintenance, self-perpetuation, and self-gratification, adding to these the mental and social reactions out of which come the religious and regulative systems. It is plain that the least derived of the mores corresponding to this classification are those which have to do with societal self-maintenance — the struggle for existence against nature and fellow-man, and for a standard of living — the most concrete and material activity of society. If then we were to call the institutions of societal self-maintenance the primary societal forms, those of marriage, the family, religion, etc., might be termed the secondary societal forms.

Recalling the above illustrations in the light of

these classifications, it becomes clearer where we are to look for evidences of rational selection. Different peoples disapprove of others' religious habitudes, their forms of marriage and of the family, their amusements, largely because these are not in the ways of the group passing judgment. Such condemnation is seldom the result of rational processes. And for corresponding reasons people approve and cling to their own ways, deeply resenting criticism of them. A number of societal forms on the line of those mentioned are practically never submitted to reason by the great majority of people, but lie in the mores, changing automatically if at all. But when one people comes to regard the economic organization of another, it usually has reason for its disapproval or may even be led to approve and adopt. There is here less of unconscious or unreasoning repugnance and prejudice. Advice is sought and taken and the prestige of superior knowledge is conceded. But compare the case as respects politics: there every man's opinion is equally good; the vote of the ignorant counts as much as that of the wise. The trouble here is that there is no immediate verification; any one can range about over the field unchecked by tests; it is like philosophy.   Why is political discussion forbidden in certain clubs?   Because it results in a clash of feeling rather than of cold intellect, and it can run to any height of passion without nearing a decisive test on fact.

For, to return to our basic proposition, it is in the field of the mores of societal self-maintenance that testing and verification are most direct and inevitable.   It is conceivable that society could live on for a long time under almost any religious form or marriage-system, without seeing it subjected to some visible and conclusive test.   But this is not so when we have to do with the institutions crystallized out of the mores of societal self-maintenance. If the mores lead to such ill success in the struggle for existence that the group in question is weakened in numbers or vitality, annihilation or subjugation is at hand.   And when man has developed the power of criticism of his own collective course, there is nothing that will provoke in him distrust of that course more speedily than economic distress.   It is upon this criterion that ministries rise and fall and policies are adopted or rejected.

It is in this field that the inadequacy of means to effect ends is most easily perceived, because the relation between means and ends is an immediate, concrete, and striking one. Hence it is here that rational selection of the mores is strongest. It is only on a primitive, irrational stage that the self-maintenance mores are allowed to check advance in the pursuit of a higher standard of living — to insist that the tree shall be felled with a shell ax rather than one of steel. Only under societal isolation, as is shown by the actual cases, could such a taboo persist in the mores. Adherence to the harmful mores becomes, even to the less developed intelligence, inexpedient; for it is plainly contrary to the most concrete and unmistakable of interests. You can persuade a savage of the inadequacy of his stone hatchet long before he can be made to see that his family system is capable of being superseded by one yielding better satisfaction to his interests.

Reason can be educated to go even farther in its selective influence on the mores of societal self-maintenance, and to reach out to the unknown-better rather than simply to discard the known-worse. Instead of taking form in blundering, unconscious effort after relief, man's mental reaction may advance to deliberate experimentation in the conscious effort to get something better. Here is where we have seen invention coming in, which is the purposeful search for the better method of procedure through the development and comparison of variations.[1] But invention has never been widely effective except in the range of the primary societal forms. In fact it has been confined largely and has become recognizedly successful and cumulative only in the development of the mechanical aids to the prosecution of the struggle for existence. No one thinks of invention in connection with the organization of the family or of the secondary societal forms in general. Any one, upon reflection, perceives that in these fields the mores move on in much the same sort of way that they have always taken. The more we know of these things the more we distrust paper constitutions and utopias, and the more commonly do we refer change to the elemental movements of the mores — though many would not use the term mores, but rather "race-character," "public opinion," "national spirit," "Zeitgeist," "Völkerpsychologie," and so on.

If the best informed and educated of men are likely to reach the conclusion that in the

---

[1] Cf. p. 46 above.

more complicated issues of societal evolution it is just as well, and probably inevitable, to "trust nature," it is because they are better aware than are those who wish to tinker and meddle, of all the complexities and difficulties attendant upon an attempt at rational selection. Dr. Maudsley,[1] for example, said of the eugenics program that he was not sure but that nature, in her blind way, could settle the issues surrounding human mating better than the human mind, through the taking of thought, could do it.   This amounts to a distrust or despair of rational selection, derived from profound realization of the difficulties and unforeseen complications involved.   It will be noted, however, that such distrust does not appear in cases where it is proposed to apply rational programs in connection with advance in the arts of life; to secure, let us say, a better means of transportation, a more satisfactory standard of value, or an extension of markets.   Here we plan confidently enough; but it is only the shallow and half-educated who do not hesitate to evolve "programs" when it comes to the more derived and less knowable and verifiable societal processes.   The truly wise stand aghast

[1] See p. 196 below.

before the tangled skein and hesitate to take hold of it; they see that the multiplicity of causes behind societal phenomena, and the consequent impossibility of foreseeing effects, are likely to vitiate any rational procedure possible to the human mind as yet evolved. They realize the precarious course that lies, in social action, between motives and consequences, where deflection is the rule, and where it would seem almost the part of omniscience to calculate times, places, and forces of impact.

The attitude of men of scientific discernment toward rational selection is about on this order: up to a certain point, where clean-cut verification stops, they proceed confidently with their projects of elimination and substitution. Beyond that they have their programs in which they believe with all the reservations; and with these programs they must go ahead on what knowledge and rational power they have, in the last analysis throwing themselves on the element of chance — that element of the unreckonable and unpredictable which surrounds on all sides the small sphere of knowledge that man's mental powers and limited energy are able to encompass. The place where this element enters is the outermost limit of the realm of rational selection, where it gives way to societal selection in its automatic, unconscious, and most elemental form.

But let us return to the distinction we have tried to make between phases of societal life where selection between variations in the mores is comparatively easy to verify and those where, on the present stage, at least, of development of the social sciences, it is difficult or impossible. Let us bear in mind that the variations in the mores are the varying ideas of men reacting to environmental conditions, and that the effectiveness of such variations is most easily tested out to every one's satisfaction (so that they can be rationally selected without much delay) in the field of societal self-maintenance. The circle-squarer or the apostle of perpetual motion is a pathological phenomenon where a crank of the same species, with a religious or political idea, happy but impossible, may attain a considerable following. In general, where the exact or scientific method can be applied rational selection between the mores is possible. But this is chiefly, if not entirely, as things now are, in that part of societal activity where men have to do with actual, concrete, natural objects rather than with each

other or with some higher power whose existence cannot be scientifically proved — that is,
where they react upon natural environment in
the effort to preserve life, or to preserve it more
satisfactorily. Self-maintenance is the primary
societal activity and its pursuit is a *conditio
sine qua non*.

Further, all other societal forms are in this
sense secondary, that their types are to a large
degree functions of the type of the forms of
self-maintenance. The hunting tribe becomes
pastoral in its type of self-maintenance; and
with this change go alterations in secondary societal forms, so that at length there is presented
a set of mores, typical all along the line, which
are referred to as the distinguishing characters
of the pastoral status. A pastoral tribe becomes
agricultural and sedentary, and presently the
inevitable attendant characteristics in the secondary societal forms appear. An analogy is
the case of a person acting " just as one would
expect " from a knowledge of his circumstances,
exhibiting a habitude typical of those circumstances. This is what is meant by the phrase
"strain toward consistency", used of the mores.
Unquestionably the secondary societal forms
come to consist with each other, and there are

doubtless minor mutual harmonizations between them which are independent of their
universal dependence upon the form taken by
the mores of societal self-maintenance.   Marriage mores and property mores consist, but it
is the mores arising out of direct reactions between the society and its environment that
form the independent variable, of which both
are functions.   The ultimate activity of society
is to preserve (feed, clothe, shelter, and protect) itself, and so the mores and institutions
that contribute to this end are in a very real
sense fundamental.   It is upon these forms as
a basis that the rest of the societal structure is
erected; and the form of the superstructure
cannot vary except in detail from the type conditioned by the character of the foundation.[1]

[1] " In general, then, when the men are too numerous for the
means of subsistence, the struggle for existence is fierce.   The
finer sentiments decline; selfishness comes out again from the
repression under which culture binds it; the social tie is loosened;
all the dark sufferings of which humanity is capable become
familiar phenomena.   Men are habituated to see distorted bodies,
harsh and frightful diseases, famine and pestilence; they find
out what depths of debasement humanity is capable of.   Hideous
crimes are perpetrated; monstrous superstitions are embraced
even by the most cultivated members of society; vices otherwise
inconceivable become common, and fester in the mass of society;
culture is lost; education dies out; the arts and sciences decline.
All this follows for the most simple and obvious of all reasons:

The perspective of societal evolution, acquired almost unconsciously by protracted contact with the concrete evidence of ethnography, fathers this conviction that facts and conditions of the economic order are the basic ones; and there are those who will agree with the position taken here as applying to primitive life, but who would reject it as untrue of conditions under a higher culture, just as they concede the importance of the influences of natural environment on the savage, but deny it in the case of civilized man. I do not care to challenge any such conviction except as it is challenged by the general tenor and upshot of my whole presentation; but I have selected my first and fullest illustrations of the basic nature of the mores of self-maintenance outside the field of ethnography or the history of primitive peoples.

because a man whose whole soul is absorbed in a struggle to get enough to eat, will give up his manners, his morals, his education, or that of his children, and will thus, step by step, withdraw from and surrender everything else in order simply to maintain existence. Indeed, it is a fact of familiar knowledge that, under the stress of misery, all the finer acquisitions and sentiments slowly but steadily perish.

"The converse of this statement, however, is true. . . . If the subsistence of men is in excess of the number of men all the opposite results are produced, for in that case the demand is in excess of the supply." Sumner, "The Challenge of Facts and Other Essays," pp. 120–121.

There is a great deal of such illustration to be gathered, but the nearest to my hand at the present time lies in certain essays about to be published, which deal in the main with our own society.[1] Since my ideas about societal evolution have developed about Sumner's treatment of the folkways, and since he was always working toward that treatment for years before it took definite shape in his mind, it is particularly appropriate to begin with him. His generalizations, as quoted, are made for the most part around discussions of American conditions.

"We are told that moral forces alone can elevate any such people again [as the Irish were several decades ago]. But it is plain that a people which has sunk below the reach of the economic forces of self-interest has certainly sunk below the reach of moral forces, and that this objection is superficial and short-sighted. What is true is that economic forces always go before moral forces. Men feel self-interest long before they feel prudence, self-control, and temperance. They lose the moral forces long before they lose the economic forces. If they can be regenerated at all, it must be first by distress appealing to self-interest and forcing recourse to some expedient

[1] Sumner, "The Challenge of Facts and Other Essays," pp. 29, 30, 51–52, 26–27, 304, 337–339. These are characteristic extracts from Sumner's treatment of societal organization. His point of view, with which I find myself in almost entire sympathy, appears again and again throughout his writings.

for relief. Emigration is certainly an economic force for the relief of Irish distress. It is a palliative only, when considered in itself, but the virtue of it is that it gives the non-emigrating population a chance to rise to a level on which the moral forces can act upon them."

"The economic forces work with moral forces and are their handmaidens, but the economic forces are far more primitive, original, and universal. The glib generalities in which we sometimes hear people talk, as if you could set moral and economic forces separate from and in antithesis to each other, and discard the one to accept and work by the other, gravely misconstrue the realities of the social order."

"The sound student of sociology can hold out to mankind, as individuals or as a race, only one hope of better and happier living. That hope lies in an enhancement of the industrial virtues and of the moral forces which thence arise. Industry, self-denial, and temperance are the laws of prosperity for men and states; without them advance in the arts and in wealth means only corruption and decay through luxury and vice. With them progress in the arts and increasing wealth are the prime conditions of an advancing civilization which is sound enough to endure. The power of the human race to-day over the conditions of prosperous and happy living are sufficient to banish poverty and misery if it were not for folly and vice. The earth does not begin to be populated up to its power to support population on the present stage of the arts; if the United States were as densely populated as the British Islands, we should have 1,000,000,000 people here. If, therefore, men were willing to set to work with energy and courage to subdue the outlying parts of the earth, all might live in plenty and prosperity. But if they insist on remaining in the slums of great cities or on the borders of an old society, and on a comparatively exhausted soil, there is no device of economist or statesman which can prevent them from falling victims to poverty and misery or from succumbing in the competition of life to those who have greater command of capital. The socialist or philanthropist who nourishes them in their situation and saves them from the distress of it is only cultivating the distress which he pretends to cure."

And, again :

"It is impossible that the man with capital and the man without capital should be equal. To affirm that they are equal would be to say that a man who has no tool can get as much food out of the ground as the man who has a spade or a plough; or that the man who has no weapon can defend himself as well against hostile beasts or hostile men as the man who has a weapon. If that were so, none of us would work any more. We work and deny ourselves to get capital just because, other things being equal, the man who has it is superior, for attaining all the ends of life, to the man who has it not. Considering the eagerness with which we all seek capital and the estimate we put upon it, either in cherishing it if we have it, or envying others who have it while we have it not, it is very strange what platitudes pass current about it in our society so soon as we begin to generalize about it. If our young people really believed some of the teachings they hear, it would not be amiss to preach them a sermon once in a while to reassure them, setting forth that it is not wicked to be rich, nay even, that it is not wicked to be richer than your neighbor."

Speaking of democracy in the Colonies, the same author says:

"No convention ever decreed it or chose it. It existed in the sense of social equality long before it was recognized and employed as a guiding principle in institutions and laws; its strength in the latter is due to the fact that it is rooted and grounded in economic facts. The current popular notion that we have democratic institutions because the men of the eighteenth century were wise enough to choose and create them is entirely erroneous. We have not made America; America has made us. There is, indeed, a constant reaction between the environment and the ideas of the people; the ideas turn into dogmas and pet notions, which in their turn are applied to the environment. What effect they have, however, except to produce confusion, error, mischief, and loss is a very serious question. The current of our age has been entirely in favor of the notion that a convention to amend the Constitution can make any kind of a state or society which we may choose as an ideal. That is a great delusion, but it is one of the leading social faiths of the present time."

Again:

"The current opinion amongst us undoubtedly is that the extension of the suffrage and the virtual transfer of the powers of government to the uneducated and non-property classes, compelling the educated and property classes, if they want to influence the government, to do so by persuading or perhaps corrupting the former, is a piece of political wisdom to which our fathers were led by philosophy and by the conviction that the doctrine of it was true and just.   There were causes for it, however, which were far more powerful than preaching, argument, and philosophy; and besides, if you will notice how hopeless it is by any argument to make headway against any current of belief which has obtained momentum in a society, you will put your faith in the current of belief and not in the power of logic or exhortation.   You will then look at the causes of the current of belief, and you will find them in the economic conditions which are controlling, at the time, the struggle for existence and the competition of life.   At the beginning of this century it would have been just exactly as impossible to put aristocratic restrictions on democracy here as it would have been at the same time to put democratic restrictions on aristocracy in England.   Now the economic circumstances of our century which have modified the struggle for existence and the competition of life have been, first, the opening of a vast extent of new land to the use and advantage of the people who had no social power of any kind; and, second, the advance in the arts.   Of the arts, those of transportation have been the most important because they have made the new land accessible; but all the other applications of the arts have been increasing man's power in the struggle for existence, and they have been most in favor of the classes which otherwise had nothing but their hands with which to carry on that struggle.   This has lessened the advantage of owning land, and it has lessened the comparative advantage of having capital over that of having only labor. An education has not now as great value to give its possessor a special advantage — a share, that is, in a limited monopoly — as it had a century ago. This is true in a still greater degree of higher education, until we come up to those cases where exceptional talent, armed with the highest training, once more wins the advantages of a natural monopoly.

"Hence it is that the great economic changes I have mentioned have produced the greatest social revolution that has ever occurred. It has raised the masses to power, has set slaves free, has given a charter of social and political power to the people who have nothing, and has forced those-who-have to get power, if they want it, by persuading and influencing those-who-have-not. All the demagogues, philosophers, and principle-brokers are trying to lead the triumphal procession and crying: 'We got it for you.' 'We are your friends.' 'It is to us that you owe it all.' On the other hand the same social revolution has undermined all social institutions and prescriptions of an aristocratic character, and they are rapidly crumbling away, even in the Old World, under the reaction from the New."

Viewed from the standpoint of what is taken to be "progress," it would seem sometimes that advance in the maintenance-mores has involved retrogression of the other mores. For example, the introduction of the factory-system seemed to throw the whole organization of society into disorder and chaos and to bring the secondary societal forms, to superficial observation, into general decline. Similar cases appear all through the last two centuries of rapid increase of economic power. But this phenomenon means no more than the falling out of adjustment of the secondary societal forms with the primary. An access of pain and want — the unmistakable sign of maladaptation — promptly ensued and forced the secondary forms into better adjustment with the primary. The former had to catch up, so to speak, with the latter.

Further evidence of a general nature, showing the basic character of the organization for societal self-maintenance, may be gathered from widely accepted generalizations of the science of society. Spencer [1] has covered this point in one of its broadest aspects in his classical comparison between the militant and industrial types of society. Militarism and industrialism are names for the two major categories of human effort by which societies are self-maintained. But the character of all the institutions of a society is determined according to which of these two prevails. This is a contention to which Spencer often recurs, and with a wealth of concrete illustration.

[1] "Principles of Sociology," II, chaps. XVII, XVIII, *et al.*

In the history of civilization there occur a number of correlations which bear witness to the basic character of the maintenance-mores. Thus an advance to an agricultural economy seems prerequisite for the development of private property in land and also of slavery; the emergence of the typical patriarchal family goes with the pastoral stage; polygamy aecompanics the prosperity and wealth accruing from an advance in the maintenance-organization. The status of women, in general low among hunting peoples, advances under a settled agricultural economy. We might include this last item in some such series as the following: the local forms of social intercourse between the sexes depend upon the local ideas of decency, chastity, etc.; these on the form of the marriage relation; this upon the position of woman; this upon her economic function; this upon the form of sex-division of labor in the industrial organization; this upon the nature of the struggle for existence; this upon conditions of the environment and of sex as they exist in nature.

Forms of diversion and of worship reflect the major occupation of the people, as, *e.g.*, the buffalo dance or the snake dance of the

Western Indians. The ritual of religion is full of reminiscences of the mores of self-maintenance of the present or past; in fact the nature of the ritual could not be what it is, did it not rest upon them as a basis. Let them change form, and the ritual will change after a time, no matter how conservative it, as a ritual, may be. The nature of governmental institutions, as the quotation from Sumner has indicated, is correlative with the local type of industrial organization. Class distinctions and the code built around them result from inequalities imposed under slavery or a competitive system whose very essence is the inequality of material reward.

International relations are based ultimately upon conditions involving self-maintenance interests. For example, a noted student of such relations has stated [1] that it is normal for a great war, such as the one now in progress in Europe, to start suddenly. If there is time for deliberation, the commercial and financial interests have an opportunity to assert themselves and to endeavor to secure some form of peaceful adaptation. They will assert themselves later on, in any case, and the final settle-

[1] Professor A. C. Coolidge, in a lecture delivered Oct. 30, 1914.

ment must include the satisfaction of the basic interests of the dominant groups.

The number of these general sociological correlations might be multiplied; they show that certain secondary societal forms simply could not come into being until the maintenance-mores had taken on such type as would admit of them or evoke them. But now I want to assemble some instances of a more concrete order, run together without much attempt at connection one with the other. Every invention that made self-maintenance easier drew after it a long string of consequences all through the societal structure; instance the early discoveries of the use of fire and of the method of conveying thought by marks; or the later discoveries, or at least introduction into the European world, of gunpowder, and of the mariner's compass; or the latest inventions, of flying-machines and of wireless communication. To develop the societal effects of any one of these in detail would be a long task; but consider the needful changes in and additions to law, which the use of air-craft must call forth, and the many results which flow out of the fact that nobody can any longer be cut off, even during an ocean trip, from his business or other interests. The societal effects of the development of trade and transportation can scarcely be measured, and certainly could not have been predicted. It seems even to have an enlivening effect upon the disposition. Graf von Götzen [1] says of a certain African tribe: "As in the case of all trading peoples the disposition of the Wassumbwa is bright and volatile, and the power of comprehension very highly developed. In association with this stands an evident capacity for adaptation to foreign customs and usages." And now a noted specialist on Homer [2] refers the Trojan War to trade rivalries and contends that the catalogue of the ships has a trade-route basis. Mathematicians show how the schools took over the Hindu-Arabic numerals from the tradesmen.[3] And clear on the other side of the world and in a quite different part of the social field, we find the development of transportation, in Jackson's time, allowing in this country a convention of commoner people (as against the intellectuals of earlier times), and this contributing strongly to the development of the characteristic American party system.

[1] "Durch Afrika von Ost nach West," p. 82.
[2] Leaf, "Troy, A Study in Homeric Geography."
[3] Smith and Karpinski, "The Hindu-Arabic Numerals," p. 136.

There are two institutions of society which are closely conjoined and correlated all through history, namely, marriage and property. The position of woman, in matrimony or out of it, varies with her economic importance, but within the institution her status is generally felt to depend upon no more than a transitory feeling unless it is steadied or rendered stable by a property guarantee. A real marriage should be founded upon permanent, positive interests. Hence the passage of property, in the form of bride-price or dowry, which distinguishes the taking of a wife of status as contrasted with the unceremonious and unguaranteed appropriation of a mere consort. This might be illustrated copiously, but a single case may serve to carry the point.

In Sumatra, where the bride-price is customary, the importance of material interests is so well recognized, that the goods paid are often overvalued in order to bring the nominal price up to the demands of the *adat* (mores). Inability to pay changes the type of the marriage; it is then by abduction. In these regions there exist side by side the marriage with bride-price (*cum manu*) and that without bride-price, called *ambel anak* (*sine manu*). In each case the whole congeries of duties, rights, inheritance-customs, divorce-settlements, etc., vary according to the manner in which property has passed or has not passed. In South Sumatra the Dutch government tried to do away with the bride-price; in 1862 it was formally put under penalty. The result is that the word for such a marriage (*jujur*) is no longer used; but the custom has been rechristened and remains. The prohibitions have, in some cases, however, coincided with the natural change (coming with a developed material civilization) toward the "parental" form (parents on an equality in rights, power to bequeath, etc.), and have thus hastened it somewhat.[1] This citation is typical of the dependence of the mores or societal self-perpetuation upon those of self-maintenance. The more primitive marriage is studied the more clearly does the investigator see that its several forms are reflections of the local economic life.[2]

[1] Wilken, "Verspreide Geschriften," II, 223–224; 230–231; 270–271, *et passim*.

[2] That the prevalence, in Australia, of the primitive form of group-marriage varies with the terms in which the struggle for existence is set, is clearly indicated by Howitt (in *Journal of the Anthropological Institute of Great Britain and Ireland*, XVIII, 33): "The most backward-standing types of social organization, having descent through the mother and an archaic communal marriage, exist in the dry and desert country; the more developed

It is plain enough that all the mores focus upon the art of getting a living as the primordial activity, and that all the rest of the code is valid only in its harmonization with self-maintenance. Says Kowalewsky,[1] speaking of certain types of settlement: "I think that every attempt to explain the types of settlement not by the need of defence and through economic circumstances, but through psychic characteristics of the nation, must finally be given up."

Let us take an illustration of this principle out of the history of a people whose mores have undergone recognizable and great change. The speediest way in which this result has been secured — whether for good or ill of the people concerned is no question here — has been by a grand alteration of the natural environment. The settlers in this country, in pursuit of their own interests, effected such a change in the environment of the aborigines. They cut down

Kamilaroi type, having descent through the mother, but a general absence of the *pirauru* marriage practice, is found in the better watered tracts which are the source of all the great rivers of East Australia; while the most developed types, having individual marriage, and in which, in almost all cases, descent is counted through the father, are found along the coasts where there is the most permanent supply of water and most food."

[1] "Ökonomische Entwickelung Europas," I, 72.

the trees, killed or drove off the game, and otherwise rendered the Indian mores in the field of societal self-maintenance no longer adequate to their ends. Progressively this process went on, until the whole societal system of the Indians broke down and they retrograded in their mores all along the line. This topic needs no extended analysis, for a little reflection upon well-known facts will show its bearings upon the subject before us.

A corollary of the contact of Indian and white man was the introduction of the horse. This altered the whole mode of the organization for self-maintenance among the Plains Indians; they became an inveterate hunting people. And with this selection of industrial mores went changes in those dealing with the rest of the societal activities. Much and extended wandering led to the development of their peculiar sign language; the position of woman declined as her economic utility decreased; in short, the societal type of the nomad-hunter was evolved in its fulness. The mores of societal self-maintenance were altered; then followed the inevitable alterations and harmonizations in the secondary societal forms.

"The Egyptians owed their power and civilization to the fact that the Nile mud so enriched the valley every season that one man's labor could produce subsistence for many. When the population increased, the power of social maintenance was not diminished but increased. When there was a great population there, using the land with very painstaking labor according to the stage of the arts, an immense surplus was produced which raised war, statecraft, fine arts, science, and religion up to a very high plane. Then they tried to satisfy the demand for men by slaves, that is, persons who contributed to the social power to their utmost, yet shared in it only under the narrowest limitations. The system, after reaching the full flower of prosperity of which it was capable, became rigid, chiefly, as it appears, because the sanction of religion was given to the traditional and stereotyped forms. Also the power of social support which lay in the fertility of the soil had been exploited to its utmost. The arts by which more product might have been won advanced only very slowly — scarcely at all. There was hardly any emigration to new land. Hence a culmination was reached, after which there must be decline and decay. The achievements of the Egyptians were made in the period when they were growing up to the measure of the chances which they possessed." [1]

I do not intend, in this place, to multiply concrete examples of the point before us. All of them would be, in a real sense, examples of

[1] Sumner, "The Challenge of Facts and Other Essays," pp. 146–147.

adaptation in the mores. Adaptation is the
characteristic outcome of evolution, and so
presupposes variation, selection, and transmission. If, therefore, adaptation can be shown,
the strong presumption forthwith emerges that
the other evolutionary factors are there. Instances of adaptation are bound to give evidence for the activity of all the factors that
lead to it. Further examples might, therefore,
conveniently be left until we come to the topic
of adaptation. There is no object in citing
at this point a mass of illustration the occasion
for which will naturally recur. The survey of
the societal system of an ethnic group, or of a
colonial (frontier) society, which can be taken
up somewhat more systematically,[1] will show
us that consistency in the mores means, at last
analysis, the consistency of all the rest of the
mores with those of the organization for societal
self-maintenance. The latter, as we shall see,
adapt themselves to the environment, which
sets the conditions of the struggle for existence; and then the former fall into line with
these latter. The code of the maintenance-
organization is the dominant chord vibrating
to the tone of the struggle with nature and

[1] P. 261 below.

fellow-man; the rest of the mores are, as it were, overtones of that chord. The latter must die away and be succeeded by others as the major chord is altered under changed conditions of the environment. Or, to use another figure, we may regard the maintenance-mores as the soil out of which the other mores derive their sustenance. Let the soil be changed in character and constituents — let it become acid instead of alkaline in reaction — and the crop must change. Let there be a retrogression or an *élan* of progress in the arts of self-maintenance, and the society will respond throughout its organization.

We have seen, now, that rational selection can take place most readily among the mores of self-maintenance, where there is the most concrete and decisive test; and then that these mores are determinative of the other mores. Converging these two lines of thought, the case for rational selection in the mores does not look as hopeless as it did. For rational selection is most effective precisely in that part of the societal organization where its operation is farthest-reaching. There are thinkers who are disposed to deny "progress" in the social field

M

as a whole; but none of them is hardy enough to assert that there has been no progress in the arts of life — in the industrial organization — along the lines of material civilization. If the positions taken above are correct, the alterations (selections) effected in the mores at the basis of the primary or fundamental societal forms will presently entail corresponding alterations in those less easily and surely modified. Thus scientific selection, applied at the point when it can be applied most readily — where its results are most easily verifiable — is bound to become indirectly effective, in a sure and natural manner, at points less accessible. This view is surely more hopeful for scientific selection than one which would deny such selection altogether because it fails so often when the effort is made to apply it directly to secondary societal forms — in fact, a candid scientist will perforce admit that in its attempted application to secondary societal forms there does not exist enough certainty, or a sufficiency of indisputable principles, to render procedure scientific at all. It is equally contrary to experience and sense to make a universal denial or a universal affirmation about the direct applicability of rational selection in societal evolution. Every one knows that it is directly effective in some cases, and altogether out of question in others. The intent of the foregoing analysis has been to find out where it is regularly effective and where not; and the conclusion is, that it is universally effective, but directly so only within a limited range.

Where rational selection is really effective, then, it enters in a manner that is not obvious but impersonal; like a force of nature, it is there in an elemental sort of way. That is why its action cannot be seen at the moment or on the surface, but must be detected, if at all, over extended periods and by a more than superficial scrutiny. That is why, viewing societal life without much perspective or insight, men think either that all is in hopeless disorder, or that anything and everything can be directly altered to suit their ephemeral or local "choice." But there can be no disorder where there is law, nor yet can there be a capricious or whimsical choice, undetermined by life-conditions.

If this be "economic determinism," it is possible to make the most of it. The objections to that doctrine seem to inhere in its application rather than in its own validity. Marx had a therapeutic plan, based upon his doctrine, for changing the mores of self-maintenance, *i.e.*, a plan of societal selection, from which most of us would, very likely, dissent; but that does not affect the truth that may be in the doctrine. It is possible to believe in something like Marxian economic determinism, and then, when it comes to a plan of action, try to help adapt society to the conditions of life as learned; it is not necessary that we should plan to alter the environment *in toto*, change human nature and other somewhat permanent elements in it, rather than accommodate ourselves, even though it is not so easy or grandiose, to life-conditions and laws. Belief in some of the socialistic doctrines, together with entire distrust in the plan for their application, does not make one a socialist any more than disbelief in the doctrine of extended state power and control renders one an anarchist with leanings towards bomb-throwing.

Acceptation of the contention set down above gives a broadly scientific basis for the survey of societal evolution and acts as a safeguard against the trickery of the plausible. It is because the mores of the maintenance-organization have moved far away from their status in earlier times that the other mores have so

altered; and since selection in the basic mores
has taken place over and over again so that
certain forms once in evidence cannot well
return again,[1] so are certain of the less basic
superseded for good and all, and no amount
of theorizing can bring them back without a
return of the basic forms to the old status.
But then, again, if the form of the struggle for
existence alters, approaching a form once
existent, all the mores take on the type corresponding, and so history may partially "repeat
itself." All the way down the scale, each
special group of mores is conditioned as to
selection within it by the nature of the wider
generic group of which it is a part, until at the
bottom all the wider groups are determined as
to their form by the maintenance-mores —
which themselves reflect the nature of the
struggle for societal self-maintenance, the form
of this struggle being itself determined by
natural conditions. Thus societal evolution,

[1] The *New York Nation* (Feb. 19, 1914, p. 177) protests against the
assumption that great men of the past, in imaginary resurrection,
would not only be "in fullest sympathy with the best movements
of the present, but that they would have discarded all the errors
and defects which clung to them during their actual lives." This
assumption of identity as between the codes of different epochs
is, like an analogy, a flimsy, though often seductive means of
appeal to the unreflecting.

whatever the inmixture of the rational, runs back ever to natural evolution. Transitional forms are at hand and the chain is unbroken.

Summing up this extended treatment of selection as a factor in societal evolution — and it is by far the hardest to render account of — we find this factor exercised at first in a form scarcely different from that displayed in the non-human organic world. Gradually, however, man comes to feel that his fate is in his own hands and to look at his own evolution as a different sort of process from that found elsewhere. Making little or no distinction, he attempts to exercise selective power over all ranges of the mores — chiefly those of others — depending for his criterion upon persuasions as to their "rightness" or "wrongness," which are largely a reflection of the code to which he is accustomed. With further enlightenment, however, some come to believe that rational procedure may be followed in selection; and with the advance of science it has become part of the mores of the civilized to try to support selective enterprises by appeal to reason. Sometimes this appeal is no more than a concession, conscious or unconscious, to modern mores, or a cloak to cover the prosecution of what unreasoning desires and interests of the old primitive order seem to demand. It is hard to find something that people want to do or do not want to have done for which so-called reasons cannot be assembled and stated sufficiently well for the purpose. But, taking societal evolution as a whole, or in long stretches, the chief way in which genuinely rational selection has been effected in the mores is through its application to the organization for societal self-maintenance. A necessary result, cousequent upon selection in this range of the mores, has been a corresponding alteration in derived or dependent ranges. The obvious practical procedure, then, if one wishes to "improve" society, is to render the means of societal self-maintenance, by judicious rational selection, better adequate to secure the ends in view. In this field the means are as readily susceptible of test as the ends are definite and concrete, and so there is some hope of attaining solid and lasting results. Then the improver might turn attention to the effort to help the secondary societal forms get into consistency or harmony with the improved primary ones. This is about as far as reason can yet go safely and securely in societal selection.   This conclusion cannot content the "world-beatifier," but it may afford a definite, though perhaps minimal hope to those whose ideals are less exalted, and who respect reason enough to wish to go a furlong with her rather than a mile, or even twain, in an emotional ecstasy.

CHAPTER VI

COUNTERSELECTION

WHILE societal selection arises out of natural selection, and while, in its early stages, amidst unmitigated violence, it is no more than a variation on the process of nature, we have seen that it takes on, as it develops, its own characteristic mode. It becomes specifically distinguishable from natural selection. It selects upon a different set of criteria. This is because it operates as between groups rather than as between individuals, and so favors social superiorities rather than biological ones. For example, in the conflicts between the Roman legions and the Germans, the social qualities of discipline, organization, etc., were plainly favored for extension and propagation over superior physical qualities. Thus does a variation on the process of nature rise up to supplant its parent stock. It is quite understandable that an observer with eye trained upon the nature-process and the resulting survival of the biologically fit, should conclude that societal selection is antagonistic to natural selection. Then if natural selection is selection *par excellence*, any selection that results in the survival of the biologically less fit must be counterselection. ✓

By way of rendering the conception of counterselection as objective as possible, I hasten to consider a list of factors in the life of modern society which are asserted to be counterselective. I do not aim at exhaustiveness of illustration, but rather at objectivity of conception about what counterselection is thought to be; nor do I try to balance the pros and cons over the listing of any particular factor as counterselective. When I have made clear what counterselection is conceived to cover, and how it should be viewed from the standpoint of societal evolution, I wish then to show how it is proposed to do away with some of it through an ambitious program of rational selection. This will throw more light upon the difficult topic of societal selection, which is the subject we have had before us now for some time.

Darwin himself [1] cites several instances of counterselection, without using the term, and

_____
[1] "Descent of Man," pp. 151 ff.

several of the authors mentioned in the Introduction to this essay have something to say along the same line. Schallmayer[1] employs the term freely and undertakes to list the social factors making for the survival of the unfit; his collection is complete enough to illustrate all the points before us.[2] I shall try to paraphrase what he says, with such comment as seems to me likely to bring his point of view out clearly.

War and military organization, first of all, though selective upon a lower stage of culture, are now prevailingly counterselective. The best are exposed to danger, while the inferior do not see the enemy; and, even among the best, the superior man physically — the strongest, the fleetest — has little better chance of survival than his inferiors. All are "food for the bullets." And even if there is no actual war, the man in the service is exposed to diseases and temptations from which the man at home is relatively protected. Again, though he may have escaped all these ills, the former has lost

[1] "Vererbung und Auslese im Lebenslauf der Völker," pp. 111 ff.

[2] Most of the matter immediately following is quoted or paraphrased from an article by the author, on "Eugenics," in the *Yale Review* for August, 1908.

several years of youth, necessary habits of
steady industry are perhaps unformed, his
chance of getting on in the world is lessened,
his marriage is delayed, his family diminished.
It might be added that he is not infrequently
demoralized for the rest of his life, and that his
children are likely thereby to suffer some dim-
inution of opportunity, or even some positive
and vital harm. Schallmayer has in mind
chiefly German conditions; America, with its
erstwhile diminutive standing-army and small
navy, has seemed to many to be profiting largely
in consequence of her beneficent isolation. But
anyone who has gained a remote conception of
the incurable injury to the quality of population
in the United States, inflicted as a consequence
of the Civil War — not to mention the economic and political burdens which we have since
borne — needs not to have it proved from
others' experience that war has a strongly
counterselective aspect.[1]

[1] Naturally there is another side to this question, one strongly
supported especially under a régime of militarism. For the
present purpose it is not necessary to balance the evidence, as
no one would be disposed to deny most of the points above.
See Schallmayer, 156 ff. Attention might be called again,
however, to the trenchant paragraphs of Spencer ("Principles of
Sociology," I, § 266) respecting the metamorphosis of societal
ideals and institutions effected by a trend toward militarism.

Again, the modern industrial organization does not call for the biologically fitter, thereby enhancing their chances in life, and so their opportunity to hand down their qualities to a numerous progeny. If a man is to be a machine, certain relatively low human activities become the object of selection. The biological ills of the industrial system have been strongly stated by many socialistic writers. At the other end of the scale of prosperity, further, the institution of property óperates to replace "natural mating" with the less impulsive type. It also ushers into domestic life not a few of the biologically unfit who are well provided with this world's goods. The way in which property is inherited operates not infrequently to limit procreation on the part of eligible parents, and to throw all the opportunities of early marriage, possession of a large family, and the like, in the way of the less fit. This was one of the counterselective factors to which Darwin gave attention, in what he said of primogeniture and entailed estates. But some critics of the French system believe that there are equally bad results accruing from laws of inheritance which involve the splitting up of estates.

Schallmayer lists also celibacy, late marriage, and the restriction of the size of families. I shall consider the last two of these rather fully as I go on. Celibacy, it is thought, is more likely to appear where marriage is delayed beyond the more impressionable years. And where it has been a policy it has often removed the pick of the young men and women from the ranks of the potential parents.[1]

One of the most unexpected of the counter-selective factors is the development of thera-peutics, hygiene, and the technique of nutri-tion. At first sight this seems to constitute the chief boon wrung by man from hostile na-ture. But, in a sense, it is a gift (to change the metaphor), in the giving of which nature might be pardoned for having been reluctant. Briefly, all these agencies, though of prevailing excel-lence, yet are counterselective in this: that they rear up to the age of maturity and of pro-creation many who formerly fell early beneath the sweep of natural selection. Precisely be-

---

[1] A number of authors, including Galton, digress while speak-ing of celibacy, generally with Spanish conditions in mind, to the activity of such an institution as the Holy Inquisition, which resolutely repressed, where it did not eliminate, any persons who had mentality or originality enough to become heretic. These left few descendants, at least in the land of oppression; and the result, for the society, was the loss of its best elements.

cause these agencies constitute, from several points of view, the most evident and important benefits of civilization, do they become unwholesome checks upon the action of the pitiless adapting agency of nature. Through the development of civilization, as has been seen, man withdraws himself from beneath the untempered sway of natural forces; here, it appears, he has probably emancipated himself almost too fully. Selection as a preliminary to mating, *i.e.*, the death before maturity of the weakly, has often been displaced and delayed until mating has been accomplished, after which time it is powerless to secure to the "fittest" of the next generation their old birthright of unburdened and unhindered development. And the old-time survival-with-little-aid as a guarantee of proper mating has as yet been replaced by no strong and controlling body of conventions or folkways.

Not to devote a disproportionate amount of space to inverse selection, but one other factor is singled out for mention: what Schallmayer calls the "individualistic tendency of humanitarianism." I take this to mean the tendency to regard a case of relief, let us say, as a relation merely of individuals, in which the society is conceded no interest. Humanitarianism in general is, of course, counterselective, for it always eases the struggle for existence and so operates — and that is perhaps its chief aim — against the elimination of those who are losing in the conflict. Carried to senseless excess, as in Spain of the late Middle Ages, it foists upon the really wholesome and vital body of society a deadly parasitic growth; beggar-migrations at one time literally invaded Spain, where un-discriminating charity was one of the surest titles to heavenly bliss. Nobody denies the social value of willingness to aid the fellow-man with discrimination; what is lost biologically is doubtless made up for socially. No one denies the social virtue, to say nothing of the personal obligation to humanity, of succoring a fellow-man in distress, whatever form the affliction may take. But that complaisance should extend to the assurance of the "right" to procreate, is another matter. "The in-dividualistic tendency" of Schallmayer is one that is blind to all those larger group-interests which at times dictate refusal of aid where the personal inclination would be to afford it. No one is fit to give charity whose sentiment over-weighs his intelligence, and whose perspective is bounded by the individual relation of giver and recipient.

We have the conception of conterselection now before us and may attack the situation, at first upon the broadest lines.

In truth, since the whole trend of civilization is to interpose barriers to the action of natural selection,[1] societal selection is logically certain to preserve those who would perish under nature. It must always be realized that societal selection operates on a different plane, demanding different standards of fitness. The only sense that there is in adjudging its results by a comparison with those of natural selection lies in the fact that the latter is the ultimate and final test. But there is a good deal of sense in doing this. In view of our knowledge that man is, at last analysis, an animal in a natural environment, there can be no doubt at all about his ultimate subjection to the laws that control organic nature. His new mode of selection is performable, as it were, only upon a scaffolding created by his civilization — only in an artificialized environment. If the scaffolding is shaken down, the environment dis-artificialized, by some catastrophe, he is again an animal

[1] P. 67 above.

N

among other animals. He cannot put more and more dependence upon the scaffolding unless he is constantly strengthening its supports. He can never bid utter defiance to natural law, ignore biological qualities altogether, cut loose once and for all from the earth. It would be dangerous indeed if societal selection were going to become unreservedly counterselective.

It is unthinkable that anything on that order will take place. In general, a strong belief in the evolutionary process fortifies one — from an *a priori* standpoint, of course — against the prediction that we are headed for ruin because we are, as some assert, breeding from the unfittest. If we had been going to ruin, we have already had enough good chances to do so. We have undoubtedly made errors and paid heavily for them; these payments in pain and loss have been due, as I see it, to the operation of selection, sometimes natural, again societal, in disposing of erratic variations that were leading to maladaptation. If we carry on our operations upon the scaffolding too violently or too near the edge, we shall undoubtedly get a fall and it will hurt; but the whole structure will not collapse thereby. What we need for safety is to know the strength of the underpinning and its carrying-power. That is the limit within which our societal selection can go its own course, whether or not it is counterselective. For safety it is necessary to restrain the antics of those whose balance and sense of caution are only rudimentarily developed. In any case we have to live on our scaffolding; even if all societal selection were counterselective, we cannot go back now to survival on the basis of biological fitness. But we do well to have an eye on the nature-processes; and those who cry out against counterselection and emphasize the need of biological fitness are performing for us the great service of insisting upon the presence of limits beyond which it is unsafe to allow variations in the mores to go.

For well-being we want to avoid the ills that attend grand-scale selection, if we can. The collapse of any section of the scaffolding of civilization shakes the whole structure, even if it does not cast it down; and those who have not been hurled into the pit of brute violence must yet suffer and sacrifice with those who have. The life and prospects of all civilized men are darkened by the war we now witness, the work of civilization is set back, and much of it will have to be done over again. The danger attending the growth of inapt variations should not be minimized or ignored — the less so as, in societal evolution, such variations may develop with a great train of consequences before they are gotten to the bar of selection.[1] If counterselection is no more than tardiness to act, on the part of societal selection, even so it is well to have the dangers of such delay set forth[2]; for if not anticipated, selection must at length be exercised by the brutal, primordial methods of general violence and destruction.

[1] Cf. p. 49 above.

[2] The following extract from the *New York Times*, of May 10, 1914, needs no comment.

"The recent agitation in the French press against the alarming spread of leprosy in Southern France reached the culminating point this week, when Clément Vautel wrote a humorous article in *Le Matin*, prophesying that leprosy would take the place of appendicitis as a popular malady and concluding that the hypochondriacs had found a new source of alarm and doctors a new source of profit. That the leprosy scare is being sufficiently advertised to become a joke paradoxically indicates its importance, while the report of the latest meeting of the Academy of Medicine did not tend in the direction of reassurance. After a warm debate the Academy voted resolutions to this effect:

"First: That leprosy be made a notifiable disease.

"Second: That lepers submit to special treatment, and even isolation, according to the gravity of the case.

"Third: That foreign lepers be forbidden to enter French territory.

"Fourth: That measures be taken to warn the public to prevent the spread of leprosy and treat existing cases more effectively.

Those who warn of counterselection are in-citers to rational selection, which, despite the difficulties attending it, we must, as civilized men, try to get the knowledge and skill and fortitude to carry out.

I have come to the conclusion, then, that part of what is called counterselection is straight societal selection; the differences between its results and those of natural selection have caught the attention, and the derived mode of societal selection has not been recognized. Cases of this order might well prove maleficent if they occurred in nature, but they do not; they are harmless enough occurring in an artificialized environment. For example, a near-sighted man would be one of the unfit under nature; but the peculiar form of human adaptation, through a materialized idea (glasses), renders him fit

" The following extract from an article in *Le Journal* indicates the state of public feeling. 'During the ignorant Middle Ages lepers were condemned to eternal solitude, to wear a hooded cloak and ring a warning bell. They were forbidden to drink and wash in public water, to enter a church or tavern, or even to speak to their fellow-men, and finally had to undergo the frightful ceremony of hearing their own death mass, receiving a handful of earth on the head, meaning that henceforth they were dead and buried.

"'Thus leprosy decreased, but the modern humane methods allow lepers to mix, trade, and even to marry. Thus leprosy spreads.'"

enough to discharge the most useful functions in a modern civilized society. The other part of what is called counterselection I take to be extreme and erratic variations in the mores which have not yet been subjected to selection at all. These are ominous as tending to expose to a heavy strain the supports (capital, labor) of civilization. I do not say that either set of cases is not counterselective under the conditions of a virgin natural environment — under such conditions even the wearing of clothes and living in houses are counterselective. What I want to show is that instances of so-called counterselection are normal phenomena of societal evolution.

No one would doubt, I suppose, that the assurance of a right to life, given by civilized societies to all who conform to what are regarded as the essential mores, is a genuine product of societal evolution. It has been pointed out [1] that the rigor of selection decreases as civilization increases; that the issue of life and death has been largely removed with the assumption by society of the duty of seeing that none of its members die of starvation or exposure, and with the limitation, within the in-group, of physical

[1] Pp. 67 ff. above.

conflict between individuals and sub-groups. It is taken to be the duty of a society to keep all its members from dying. This is why societal selection, except in its rudest forms, cannot hope for the clean-cut results reached by a selection issuing in survival or non-survival. But the very fact that all members of a society are to be regarded as equally entitled to protection by it constitutes a real departure from nature's norms, for there is no distinction made between biologically fit and unfit. Perhaps we cannot have genuine counterselection unless the fit are put under disadvantages in favor of the unfit; but a selection which leaves fit and unfit upon an approximate equality in the competition of life is far enough from natural selection to deserve the counter designation. I shall not seek to emphasize the distinction between these two forms, as it would seem to serve no useful purpose in the present instance. But it is clear that such counterselection is normal in societal evolution.

So much concerning the right to live; now as to the opportunity to propagate. The fit in nature are recognized by the fact that they are numerous. Fewer die and more are born. In society the death-rate is modified as I have described. No doubt the success in reducing the death-rate — in increasing the average length of life — has redounded somewhat more to the advantage of certain parts of the population than of others; but certainly there is a strong tendency toward the dissemination of achievements along this line. In a general way the various groups in a society do not spread so very widely in this respect. But there yet remains the birth-rate. In general it declines with the advance of standards of living, so that we even hear of race-suicide. But there is a spread between the groups in this case; we find that the count of offspring per pair varies widely among the different groups in a modern society, being correlated as it is so closely with the standard of living. Certain classes are said to be barely or scarcely reproducing themselves; while over against them stand what one author has called the "swarming, spawning millions."

Leaving out of account, for the moment, the deliberate restriction of offspring, we have Galton's calculation [1] to demonstrate to us that anything which delays marriage acts strongly to reduce the number of those who put it off,

[1] "Hereditary Genius" (1870 ed.), pp. 353–356.

relatively to that of those who do not. If
M and N are each 22 years of age, and M and
all his descendants marry at once, while N and
his descendants marry at 33, the proportion of
mature M's to N's will be, in 300 years, as 26 is
to 1.   Elsewhere [1] Galton states that the fertility of mothers married at the ages of 17, 22,
27, and 32 is as 6, 5, 4, and 3.   He concludes
that a "heavy doom" hangs over the group
that practices late marriage.   But it is a fact
of common knowledge that there is a great dis-
parity as to the age of marriage between various sub-groups in a modern society.   Hence the
lowering of the birth-rate has not affected society evenly throughout its constituent groups;
it has declined much for some and not much,
if at all, for others.   That means that some
groups are increasing faster than others and
will presently constitute a preponderant fraction of the whole.   If such increasing constituents are the unfit, evidently we have a clear
case for the counterselectionist.

Perhaps the alarmist would have some trouble
proving that the less fecund classes are the
biologically fitter, but we can let him, in this
case, carry his assertion to the plane of societal

[1] "Human Faculty," pp. 321, 323.

evolution.   He can say that those classes which the society itself recognizes as of superior value to it — the highly trained and efficient — produce more slowly than those who have little more than their bodily strength to put into the social stock of power.

Now this is doubtless true, and it is significant.   It means a slower development of the highly equipped.   It is very likely an unfavorable variation.   In any case those who exhibit it do disappear by what has been called "race-suicide," after a while.   But their places are filled and the process goes on.   It is, in fact, entirely normal in societal evolution.   It will be recalled that in the Law of Population, as applying to man, there is the element of the standard of living, with which varies inversely the growth of numbers.[1]   The rise in this standard means limitation of numbers.   But its presence in a high power is characteristic of those who know and who foresee; and these are the trained and efficient.   With a high or a higher standard of living in view they marry later and, possessing the continence or the knowledge of means of restriction, have smaller families.   It is a case of the sensitive part of

[1] P. 24 above.

a society performing an adaptation, in part for itself, but, more widely viewed, also for the society as a whole; for here we see a society adapting itself to its available means of subsistence on a given stage of the arts.

These are two of the widest aspects of counterselection, since they have to do with the fundamental functions of living and of procreating. Doubtless all of the factors listed at the beginning of the chapter have contributed to the survival of biologically unfit individuals or even sub-groups; but from the standpoint of the general evolution of societies they seem to me normal, either representing forms of adaptation, on the part of the society in whose mores they are, to the environment in which they exist; or constituting, it may be, extreme and erratic variations. Celibacy, certainly ecclesiastical celibacy, belongs to the latter category, and in its pronounced form was presently selected away. In fact it was never enforcible, running as it did athwart the path of one of the strongest natural impulses; some of the facts are collected by Mr. Henry C. Lea, in his "Historical Sketch of Sacerdotal Celibacy in the Christian Church." And yet even caste-celibacy is a natural enough offshoot of the

differentiation of function in society — one of
the first essentials in societal adaptation.

Recurring now to the rest of the counterselective factors listed by Schallmayer, we come first
to war and the military organization. But we
see that, even though it discriminates against
the biologically fit, the warfare of civilized
peoples must be as it is: the evolution of war
has led from individual conflict to the collision
of larger and larger groups. It is now no longer
man to man, but army to army; it is not even
ship to ship, but fleet to fleet. It requires
special training of great numbers; and it cannot be said that those ineligible as actual com-
batants do not help by their losses to pay its
price. Out of this situation, which is itself
normal enough in societal evolution, arise ills
which rational selection is engaged in lessening
as it can. Movements toward peace and peaceful settlement of collisions of interest are not
without their significance.

In the industrial organization the same shift
from the individual to the group, characteristic of societal evolution, is patent. The unit-
man does not count unless he allies himself
with other units having common interests, as
in trade-unions, and then, through industrial

war, struggles for their realization. Discomfort and pain call for such struggle toward adaptation, and it is taking place. Again, whatever may be said of "natural mating," marriage and property have always been closely linked together and the union has rendered the family more stable, which was necessary in the interest of the society. Primogeniture is often favorable to a weak first son, it may be; but it developed along with the societal evolution of property, and seems to have no more ill effect than the system of sharing property among all the children. Societal evolution, I repeat, is the evolution of society and its institutions, not of a series of individuals.

As respects the counterselective aspect of medicine and hygiene, in that they preserve the physically weak until they can marry and pass on their unfitnesses, it is perhaps superfluous to recall yet again the difference in mode and plane of the societal type of evolution.[1]

[1] The following report of an address by Professor Sedgwick (*Brooklyn Eagle*, quoted in *New Haven Journal-Courier* for Dec. 22, 1913) illustrates several of the points before us, but particularly the efficiency of automatic selection operating through the self-interest of individuals and groups.

"There is a pretty deep significance in the idea presented by Prof. W. T. Sedgwick before the life insurance presidents, that it is much more important for society to save the man of 40,

The sickly mature have often contributed to the strengthening and adjustment of the institutions of society, though by procreation they may have subtracted something from its biological soundness. The question always is the same: What will the scaffolding stand? Is the advantage worth the danger incurred?

This danger should never be minimized, but it is often not so great as it seems at first sight. Attention is called to the fact, as an instance of

the trained wealth-producer, than to save babies. The professor does not, of course, oppose or deprecate efforts to reduce infant mortality, but he favors and exalts such community experimentation as will prolong the period of efficiency in human life, after efficiency has been proven.

"If babies are kept from dying, it is manifest enough to the economist that a fairly large proportion of those saved are congenitally inefficient; that the interference with the law of the survival of the fittest, justified as it is by all humanitarians, has disadvantages as well as advantages. It may and often does burden the future with incompetents to be supported or partly supported by the community or by individuals.

"As Professor Sedgwick says, the man of 40, skilled at his trade, a steady wage-earner, has passed the test of 'survival of the fittest.' He is one of the fit who have survived. What can the community do to prolong his productive energy?

"Clearly secure tenure of a job, adequate wages, fair hours and encouragement to use leisure intelligently must be considered. The need of quick and scientific attention to any sickness is likewise apparent. Ventilation in working rooms and in homes is important. Other details will suggest themselves to keen minds. The self-interest of the life insurance companies, enlisted to save the man of 40, may be of immense advantage to the human race."

counterselection, due to advance in medical science, that women need more and more care at the time of child-birth; that even domestic animals are coming to require it. Very well: they can get it, in our artificialized environment, just as the near-sighted person can get glasses. Similarly with other such needs. Civilization, to be sure, has led to the crowding of people together, and this has been attended by the development of large-scale maladies; but the general course of societal evolution has led to cities and slums, and it could not have been otherwise. Thus even the disease is normal — and, it may be added, so is the crime. We adapt ourselves to these conditions in various ways impossible under nature. There is constant call for more knowledge, especially knowledge that will contribute to societal welfare, and for rational selection operating in its light.

The case is similar with humanitarianism. It is plainly counterselective, viewed from the standpoint of the zoölogist, even in its least pretentious manifestations, as in the protection of life alluded to above. But, on the plane of societal evolution, it is the society's interests, always, that are paramount; and those interests are conserved, first, by the development of antagonistic coöperation and, later on, by closer forms of integration. Extremes of short-sighted humanitarianism are variations not yet pruned away by selection.

Hence, as it seems to me, there is not much truth in what some people appear to believe, *viz.*, that what they call counterselection is something abnormal; that conditions allowing of it ought not to have been permitted to develop. Their development was inevitable on the plane of societal evolution. But I do not wish at all to deride the criticisms of our social order, implied or expressed. I have said that before. We are all organic beings, and as such, subject ultimately to selection. Every one of the counterselective factors we have considered, and others too, are, in their extreme variations, menacing to our well-being and to that of society. Their presence calls for tirelessness in the acquisition of knowledge and unabated vigilance in the exercise of rational selection based upon it. And we have not been utterly slothful and torpid in doing this. Take one illustration: the case of humanitarianism. We would never duplicate Spain's performances in the matter of the beggars' crusades. Charity organizations, manned by a personnel which includes cool heads as well as warm hearts, are becoming agencies of intimidation to the sturdy rogue; for him they are turning into houses of correction. Here is hope which we must not forget when we encounter the mores of sentimentalism of which I have spoken elsewhere.[1]

In this chapter, which is one of digression and illustration rather than in the direct line of argument, the topic of eugenics can scarcely be passed over. Counterselection consists in good part in the elimination of the biologically fit from their proportionate share in the procreation of the next generations; but eugenics is concerned with having future generations of the highest possible quality. Hence it is bound to run foul of counterselection. And the fact that it is a special and aspiring program of rational selection renders its examination, now that we have surveyed the general case, the more useful and appropriate. An amount of indiscriminate effort at "social uplift" now goes under the name of eugenics. Imbecile pretentiousness and yearning sentimentality have seized upon the term and are rapidly and glibly laying it waste. In what I have to say

[1] Pp. 69–70 above.

here of eugenics, I always refer to it as understood and expounded by its founder, Francis Galton.

Galton wanted to secure, in time, the rational breeding of human beings, that is, rational human mating within the marriage institution. Eugenics has to do with "all influences that improve the inborn qualities of a race; also with those that develop them to the utmost advantage." [1] As in the essay quoted, attention is here confined to "the improvement of the inborn qualities, or stock, of some one human population." "The aim of Eugenics," says Galton,[2] "is to bring as many influences as can be reasonably employed, to cause the useful classes in the community to contribute *more* than their proportion to the next generation." This is to be done through investigation followed by popularization of results; by these means Galton hoped at length — after generations — to introduce eugenics "into the national conscience like a new religion." "I see no impossibility," he says,[3] "in Eugenics becoming a religious dogma among mankind, but its details must first be worked out sedu-

[1] Galton, "Essays in Eugenics," p. 35.
[2] *Ibid.*, p. 38.   [3] *Ibid.*, p. 43.

lously in the study. Over-zeal leading to hasty action would do harm, by holding out expectations of a new golden age, which will certainly be falsified and cause the science to be discredited. The first and main point is to secure the general intellectual acceptance of Eugenics as a hopeful and most important study. Then let its principles work into the heart of the nation, who will gradually give practical effect to them in ways that we may not wholly foresee."

Here is a plan of rational selection which possesses the unique character of an attempt to secure better societal adaptation through rational action on heredity.[1] It stands out for this reason among the many tentatives which aim at alteration of the environment. I do not intend to describe it here in detail; the works of Galton [2] and others have set it before the world. I only wish to view eugenics from the standpoint of what we have learned about the nature and limits of such rational selection.

I have spoken of the program of eugenics as ambitious. In a former connection [3] we

[1] Pointed out in conversation by Professor J. E. Cutler.

[2] "Essays in Eugenics." A summary occurs in several of the contributions to: "Eugenics; Twelve University Lectures"; collected by Lucy James Wilson.

[3] Pp. 91–92 above.

have seen that the mores are resistive of change in proportion as they are of long standing and are inveterate. But the plan here is to alter, not the ephemeral folkways of fashion, nor yet a method of industry; not a type of amusement, nor yet a local political form; it aims at a revolutionary change in the mores which surround one of the most vital functions of individual or society — that of self-perpetuation. It is hoped that sedulous labor in the study will here work out the details of what may then become a religious dogma. Certainly this enterprise is ambitious — so ambitious that an authority like Dr. Maudsley pronounces,[1] after listening to the proposals: "I am not sure but that Nature, in its own blind, impulsive way, does not manage things better than we can by any light of reason or by any rules which we can at present lay down." Distrust of the human agency, that is, of any form of selection except the automatic, assails one who has any conception of the difficulties involved in this project. For into this whole matter comes the eternal question of compensation and counterbalance, before whose unexpected eventualities the human mind stands

[1] "Sociological Papers," 1904, p. 54.

abashed, and is fain to resign the issue — as yet — to chance, or Nature, or God.

The only general answer to such a counsel of discouragement is that, for good or ill, man has been interfering with his own destiny now for some ages. It has become a sort of "natural law" that he shall. Looked at in the broadest way, all human civilization, as we have seen, has been the gradual accumulation of maxims and methods calculated to withdraw men from beneath the undisputed sway of nature-forces. From the invention of the hollowed log to the construction of an ocean greyhound, from primitive barter to modern exchange-operations, man has been putting his forces to better use or economizing them, in his endeavor to live, and then to live on a better scale. One may harbor reasonable doubts as to whether we are any happier for being civilized; he may feel that for every increase in the numerator of the fraction there has taken place an equal or greater accession to the denominator; a scrutiny of the counterselective factors, several of which are the direct outcome of "conquests over nature," upon which the race has plumed itself, may suggest to one that we have been all the time tugging at our own boot-straps;

but notwithstanding all this, it is evident that it is man's earthly destiny, under some powerful natural constraint, to persist in setting up his reason against "natural law." He will continue to do so. Hence the objections based upon misgivings and fear, however strong theoretically, fall out of practical reckoning; that a thing looks doubtful or impossible has never seriously deterred man from attempting it if he has wanted to — and sometimes his assault has not been a failure.

This may be called a pious hope rather than an encouraging observation. To particularize a little more, it may be recalled that while certain ranges of the mores, where verification and demonstration are most readily secured, are more amenable to rational selection than others, the former do not include the mores of societal self-perpetuation. Sex-passion is a force that does not easily submit to be bridled except under the compulsion of restraints evolved long ago. Galton tacitly admits this point when he goes back for his cases of control of sex-passion to primitive peoples and the childhood of societal organization. He cites as "restrictions on marriage": monogamy, endogamy, exogamy, the Australian marriages, taboo, prohibited degrees, and celibacy. The implication is that sex-passion has been restrained and so can be again; and the agency is to be rational selection, as proposed in the eugenics program. But let any one explain and demonstrate how the above restrictions, or any one of them, arose out of rational selection. The origin of all such forms is lost in the past; but they must one and all have arisen before man was capable of conscious rational selection.[1] They go back to the ruder forms of selection practiced in the elder ages; it was those ruder forms, more nearly like those of nature, that alone had the power to make broad and sweeping selections such as are represented in Galton's list — and it is to such a list, I repeat, that he must have recourse to find a selection analogous in sweeping quality to that proposed in the program of eugenics.

Hence I cannot see the force of these supporting instances on an issue presentable to civilized peoples only, and which must involve rational selection. Still less can I share Galton's hope that eugenics can be introduced "into the national conscience like a new religion." It is after such selection only as has produced the

[1] Cf. p. 64 above.

taboo, uncriticizable and imperative, that the mores can get the sanction of religion — and by religion I mean "old" religion, not "new," that is, an unreasoning fear of the unknown and supernatural, not a respect for the known and demonstrated. I mean fear of ghosts and goblins, not the gentle fright that is aroused by a logical or experimental demonstration. If one considers the origin and nature of religious dogmas, he will hesitate to believe, with Galton, that eugenics, after sedulous work in the study, will emerge as such. Any sociologist knows that to talk of a "new religion" is simply playing with words; that there is no real religion that does not rest upon unreasoning fear of the unknown. Only so does it exercise, for the masses of mankind, any compulsion on conduct. If any one thinks he can introduce a religion among the masses by the use of statistical tables or microscopic sections, together with conclusions based thereon, he is not rational enough to be talking about sociological subjects at all.

I do not believe that Galton meant literally what he said about eugenics becoming a religion; his brief utterance on "Eugenics as a Factor in Religion"[1] bears me out. He seems

[1] "Essays in Eugenics," pp. 68–70.

to mean by a "new religion" a new ethics, that is, a new set of mores. But people less wise than Galton have meant literally what Galton said, and the statement has seemed to me deserving of a challenge.

. It is evident that the application of what we found to be true about rational selection to this case of proposed rational selection cannot leave us sanguine as to its realization. But a conclusion that the positive program of eugenics stands small chance of being carried out under conditions as they exist in human society does not exclude the possibility of partial realization. It is a waste of time to deduce and to rhapsodize about the details of a scheme, supposing it about to be realized — on the principle that "you cannot state the consequences of what never happened" — but there is in the eugenics program a great idea which we can try to realize, in so far as the limits within which our action is placed will allow, and thus add to the sum of our civilization.[1] It is the strength of resistance encountered in the mores, of course, which measures the hope of realization of any such program, in whole or in part; and of partial measures it is the

[1] Cf. p. 19 above.

one which is capable of concrete demonstration that meets less resistance than the rest. Those measures of regulation stand the best chance of success which show a close relation to the concrete interests of individuals and groups, which seem to touch most closely upon local spheres of self-maintenance and self-realization. This is a point of view with which we are now familiar. And, finally, it is easier to demonstrate the need of cutting off certain practices, whose results have sometimes been verified over and over again, than it is to enlist sympathy for some positive project, however reasonable, which has never been tried. Men have been taught by sorry experience to mistrust reason and logic. It is easier, then, to enact a proscription than a prescription. Regulations of conduct, from the Tables of the Law forward and back, have been couched in the negative; "Thou shalt not" has always been a more clean-cut and cogent form, less exposed to misconception, than "Thou shalt." Negative regulations lend themselves less readily to interpretation or cavil, and so inspire more confidence that they will be observed.

In conformity with these conditions it is that form of eugenics that is called "negative eugenics" which shows some prospect of being able to introduce rational selection into the field. Non-eugenic practices can be crippled, if they are not eradicated, and may be held within bounds, narrower or wider. In any society the wiser and more powerful are constantly handing down the ways of their group to the more ignorant and less powerful, and by virtue of their superior intelligence they are able, through the control of the regulative organization, to set up interdictions which, if they do not have the force of taboos sanctioned by religious fear, have at least the power of the state behind them. Such regulations, as I have said, are effective in proportion as they are negative and concrete in form, like the primitive taboo; contrast the laconic, "Thou shalt not kill" with the diffuse, "Thou shalt practice eugenics." This latter sort will not do; societal control must be more rude and peremptory and more definite. It must make utterance more as follows: "Thou shalt not, being an idiot, marry and beget thine own kind." Eugenics legislation can turn resolutely to the heavy-handed prohibition of the grosser, the more obvious and undeniable phenomena of counterselection. Very likely an almost general consent could be gained to the prohibition of the union of imbeciles; in fact, laws now exist forbidding it, and they are not so foreign to the feelings and prejudices of the masses as to be dead letter.

But when we try to prohibit more than the obvious cases, we encounter, again, the limits set by the folkways; for the latter have a certain consistency among themselves, and an attempted alteration of certain conventions may wreck on the opposition of certain others, which are apparently, at times, only rather distantly correlated with the ones assailed. Recent laws about sterilization horrify many sensible people; and to justify such a horror there is always a reservoir of argumentation, illustration, and interpretation to be drawn upon. As the gross and obvious is left behind and an attempt is made, for example, to insist upon a physical examination of those who propose to contract matrimony, the ranks of the objectors (and so, of course, of the evaders) fill up. The prospects of the laws about sterilization and the health-certificate for marriage do not seem bright. The lack of rational basis in them is not accountable for this; it is not upon that criterion that they are opposed or evaded. The trouble is that they are not consistent with the body of the code. Here are again the limits of a program of rational selection aiming at eugenic regulation, as set down in the mores. These limits are, after all, very narrow; and experience shows that they are not overcome when once the law is inscribed in the statute book.

If the reformers wish to do what can be done instead of dreaming about a world-beatification that has no prospect of occurring, there is yet another way to take hold of the issue. They may not be able to enlist the enthusiasm of the ordinary man for the welfare of the tenth or fifteenth generation from now; but if they can visualize for that man the fact that he is paying, here and now, out of his own fund for self-maintenance, for the support of those who should never have been born, they can touch him upon a very sensitive interest. If every consumer realized that he himself is paying the customs dues under a protective tariff — if the tax were direct instead of indirect — there might conceivably be a totally different alignment of opinion based upon a shifted conception of interest. So here. This is where any campaign of education finds its strength: it engenders more rationality in selection. This point scarcely needs to be labored here, as the general case appears in the next chapter. That which touches upon the struggle for a living, and for a standard of living, is most readily apprehended, and here action is closer on the heels of understanding than it is elsewhere in the field.

I have taken the case of eugenics to illustrate a program of rational selection directed against a case of counterselection. It is an extreme case, proposing as it does an alteration of deep-seated mores surrounding the union of the sexes and an establishment of new ways in that which has to do with the compelling passion that insures society's self-perpetuation. The case is at its hardest; but what is true of it, is true, in its degree, of any scheme of the sort.

The subject of societal selection is by far the most difficult of the topics which we have set out to survey under societal evolution. I do not see how it is possible to cover it exhaustively, even in a large treatise; but I have tried to set down its essential phases, and I think it has been shown to be a massive and elemental factor, fit to take its place in an evolutionary formula.

To those who accept the essentials of the last chapters, societal selection cannot, at any rate, remain merely two vocables, used more or less at random to connote an uncertain something in the life of society vaguely perceived to be more or less analogous to Darwinian selection in the organic world.

CHAPTER VII

IMPLICIT in the preceding pages lies the assumption that the folkways are not a matter of one generation or of one group alone; they are not localized in time or in place; they can be transmitted. In examining variation and selection there has been some need of modification in the connotation of those terms as used in organic evolution, to suit them to the new medium where it was designed to employ them. Thus variation in societal evolution has nothing directly to do with the "ids" and "determinants" of the germ-cell (to use Weismann's conception by way of illustration); and selection is not always — among civilized peoples, not often — effected through the actual extinction of life. These terms are generalized, without, however, as it seems to me, destroying their validity as truly evolutionary factors. Similarly in regard to transmission; it can hardly be maintained that the agency involved in societal evolution seems, at first sight,

closely akin to heredity. In a sense, variation
is always variation; and elimination, while
diverse in degree and mode, is always elimination. Variation through "ways" might even
be looked upon as a result of organic variation,
e.g., in race-character; and societal selection is
seen to have had its root in natural selection.
But there are doubtless many who would say
that there is nothing in societal evolution corresponding closely enough with heredity in natural evolution to lend to an argument for societal
evolution the weight of more than a vague analogy. If this were self-evident, there could be no
object in going on in this direction, for "reasoning from analogy" is out of date in science.

It would be easier to range the transmission
of the folkways under an evolutionary formula
if one were disposed to connect it directly, at
least in some degree, with heredity. This,
since the mores are always acquired characters,
could be done by assuming forthwith the inheritance of acquired characters. It would
thus be explained, for example, that certain of
the sex-mores (modesty) are really inherited;
that various ways of women, regarded as
peculiarly feminine (submissiveness), are the
inherited results of oppression on the part of

man; and so on. But there is too much evidence against the theory that acquired characters are inherited to render progress under its auspices other than precarious; whereas if the case for transmission is rendered tenable on the opposite hypothesis — namely, that at the basis of Weismannism — then any possible return to the alternative theory could but strengthen the case.

It is not proposed to assert sweepingly that heredity does not enter into the transmission of the folkways at all. To take the instance alluded to in the last paragraph: women must have differed in the degree of their submission to man, and this must have been due at bottom to differences of temperament, etc., which were not acquired characters at all, and which would doubtless be transmissible through the germ-plasm. Under the conditions prevailing upon certain primitive stages the submissive type would be the only one suited to the environment, and so the qualities of patience and endurance by which the female sex is characterized [1] would become distinctions favorable

---

[1] Ellis, H., "Man and Woman"; Campbell, H., "Differences in the Nervous Organization of Man and Woman"; Sumner, "Folkways," Chap. IX.

to persistence. Hence the ways corresponding to these qualities would tend to characterize the sex. This does not mean that the sex mores themselves, or any other mores, are transmitted by heredity; as in the case of diseases, where the theory is that it is not the disease but the predisposition which is inherited, so here: it is not the mores themselves, but the tendency toward submission — or some other psycho-physical quality — which is carried by heredity. Thus heredity may well be connected indirectly with the transmission of the mores; but it is not the characteristic agency.

In general it may be said that the mores have to be learned; they are acquired characters of the groups practicing them. One of the clearest typical cases of the mores is language[1]; but, as everybody knows, although the various human groups have used language for ages, every infant (or every congenitally deaf person, upon restoration of his powers) has to learn it anew. Again, no child is born with "manners"; children are born unmannerly, and it takes the whole force of the societal environment, including parental discipline, to coerce the

[1] Sumner, "Folkways," §§ 135-141.

young life into the mores of its time and place. The mores have to be learned just as the rules of a game have to be learned; they are the rules of the game for life in society. And as such they constitute constant limitation upon "natural" action, limitation called for by the presence of others.

The factor in societal evolution corresponding to heredity in organic evolution is tradition; and the agency of transmission is the nervous system by way of its various "senses" rather than the germ-plasm. The organs of transmission are the eye, ear, tongue, etc., and not those of sex. The term tradition, like variation and selection, is taken in the broad sense. Variation in nature causes the offspring to differ from the parents and from one another; variation in the folkways causes those of one period (or place) to differ from their predecessors and to some extent among themselves. It is the vital fact at the bottom of change. Heredity in nature causes the offspring to resemble or repeat the present type; tradition in societal evolution causes the mores of one period to repeat those of the preceding period. Each is a stringent conservator. Variation means diversity; heredity and tradition mean the preservation of type. If there were no force of heredity or tradition, there could be no system or classification of natural or of societal forms; the creation hypothesis would be the only tenable one, for there could be no basis for a theory of descent. If there were no variation, all of nature and all human institutions would show a monotony as of the desert sand. Heredity and tradition allow respectively of the accumulation of organic or societal variations through repeated selection, extending over generations, in this or that direction. In short, what one can say of the general effects of heredity in the organic realm he can say of tradition in the field of the folkways. That the transmission is in the one case by way of the organs of sex and the germ-plasm, and in the other through the action of the vocal cords, the auditory nerves, etc., would seem to be of small moment in comparison with the essential identity in the functions discharged. Tradition is, in a sense and if such comparison were profitable, more conservative than heredity. There is in the content of tradition an invariability which could not exist if it were a dual composite as is the constitution of the germ-plasm. Here we must recall certain essential qualities of the mores which we have hitherto viewed from another angle. Tradition always looks to the folkways as constituting the matter to be transmitted. But the folkways, after the concurrence in their practice has been established, come to include a judgment that they conduce to societal and, indeed, individual welfare. This is where they come properly to be called the mores.[1] They become the prosperity-policy of the group, and the young are reared up under their sway, looking to the older as the repositories of precedent and convention. But presently the older die, and in conformity with the ideas of the time, they become beings of a higher power toward whom the living owe duty, and whose will they do not wish to cross. The sanction of ghost-fear is thus extended to the mores, which, as the prosperity-policy of the group, have already taken on a stereotyped character.[2] They thus become in an even higher degree "uniform, universal in a group, imperative, invariable. As time goes on, they become more and more arbitrary, positive, and imperative. If asked why they act in a certain way in certain cases, primitive people always

[1] Sumner, "Folkways," § 2.   [2] Pp. 57–58 above.

answer that it is because they and their ancestors always have done so." [1] Thus the
transmission of the mores comes to be a process
embodying the greatest conservatism and the
least likelihood of change. This situation represents an adaptation of society to life-conditions; it would seem that because of the
rapidity of succession of variations there is
need of an intensely conservative force (like
ethnocentrism or religion) to preserve a certain
balance and poise in the evolutionary movement.

Transmission of the mores takes place through
the agency of imitation or of inculcation;
through one or the other according as the initiative is taken by the receiving or the giving
party respectively. Inculcation includes education in its broadest sense; but since that

---

[1] Sumner, "Folkways," § 1. Proverbs and other less concentrated forms of folk-wisdom, once tested and found good —
and never, it should be added, to be set aside without examination,
as foolish or childish — constitute the primitive mode of accumulation of experience. For example, the Yorubas "have an extraordinary number of proverbial sayings and regard a knowledge of
them as a proof of great wisdom, whence the saying, 'A counsellor
who understands proverbs soon sets matters right.' They are
in constant use, and another saying runs, 'A proverb is the horse
of conversation. When the conversation droops a proverb revives
it. Proverbs and conversation follow each other.'" Ellis, A. B.,
"Yoruba-Speaking Peoples," p. 218.

term implies in general usage a certain, let us say, protective attitude taken by the educator (as toward the young), the broader and more colorless designation is chosen. Acculturation is the process by which one group or people learns from another, whether the culture or civilization be gotten by imitation or by inculcation. As there must be contact, acculturation is sometimes ascribed to "contagion."

This is a convenient point to set forth the distinction between acculturation and parallelism. Many a phenomenon has been referred to the former which is plainly a case of the latter; this is simply because the explanation was easier and more obvious in the one way than in the other. Parallelisms are fine cases of adaptation [1] resulting from selection; they occur where groups which could not have been in contact allowing of acculturation — that is, the transfer of culture, or, more broadly speaking, of the mores — display identical or closely similar societal forms. It is, of course, easier to say forthwith that group A must have given group B such and such societal forms through such and such probable or possible means of transmission, than it is painfully to investigate

[1] P. 257 below.

cases of similarity, rigidly exclude clever guesswork as to the means of transmission, and develop the argument along the line of the evolutionary reasoning. In the present chapter, which has to do with transmission, no account is to be rendered of parallelism.

Recurring, then, to tradition, we find its two sub-forms as above distinguished to be of very unequal effectiveness. In the positive and conscious form of transmission, inculcation or education, the aim is to pass over to the recipient all the good, i.e., successful mores; while imitation unconsciously transmits all the mores, even those about to be eliminated. The boy, let us say, has his father's virtues held before him that he may acquire them; meanwhile he quickly and almost unconsciously imitates his ideal in other respects. The subject race is to be lifted up by the inculcation of selected European mores; but in the meantime it acquires through imitation the peculiar vices of civilization. It is plainly to be seen that of those two forms of tradition, or transmission of the mores, imitation is by far the more effective, natural, and elemental; compared with it the other agency is artificial and late of development. In fact inculcation fills in transmission something of the position that rational selection holds in selection : it represents the inmixture of the human mind, acting consciously and with some end in view. Education is gravid with theory and its aims vary from time to time; imitation, being unconscious, has no theory or aims to pursue or to alter, and so goes its course with the steady, consistent, and relentless urge of a natural force. Education or inculcation of the mores implies, in short, an antecedent display of rational selection over the mores; and so its programs are open to all the misgivings which, as has been seen, cluster about all programs of rational selection.

It is always useful to consider societal processes in comparison with the natural ones, for this renders the conception of the former more objective. Thus, while transmission of organic characters is always from parents to progeny, from the older generation to the younger, in the case of the mores this universal does not hold. Education generally has reference to transmission with the younger as recipients, and the faculty of imitation is undoubtedly stronger and is exercised to a greater degree by the young; but there are, none the less, many cases of transmission in the reverse direction (*e.g.*, parents learning from their children), and acculturation within or between groups of the mature is common. Thus arise in societal evolution complications of an order not represented in the organic field. We are dealing with societies rather than individuals or pairs — and in so doing, it may be added, transmission is found to be chiefly an intra-group rather than an inter-group process, for the checks upon transmission, previously alluded to,[1] such as ethnocentrism, are operative for the most part between groups.

By way of getting the essential features of tradition before us, let us consider the simplest case, *viz.*, the intra-group transmission of the mores from the older generation to the immature. This must include both forms of tradition as sketched above, but will assist in their further distinction. There is here a virtual cancellation of the factor of diverse heredity in its stronger manifestations, as, for example, between races, the assumption being that within the groups there exists the basis of a common heredity. Such a group would be represented well enough, though not by any means

[1] Pp. 57–62 above.

perfectly, by England, France, or the northern states of this country; and still more perfectly by a smaller, in some way segregated community, as, for instance, New England in the Colonial times.

In organic evolution there is a "biogenetic law" or law of recapitulation, which has thrown unexpected light upon the evolutionary process and has lent peculiar strength to its demonstration. This law asserts that "ontogeny repeats phylogeny," or that the life-history of the individual is an abbreviated repetition of that of the race. Strictly speaking, the life of the individual referred to means embryonic life; but there seems to be little doubt but that the life of the child after birth carries on its recapitulation into further phases. However that may be, the child speedily develops another kind of summary of preceding racial phases; for he begins those mental reactions which, as was seen early in this work, represent man's characteristic mode. If, now, a societal law were to be formulated on the lines of the biogenetic law, it might run: The life history of the generation rising to maturity is an abbreviated repetition of the societal evolution of the group. No "argument from analogy" is sought; but it is desirable to consider the last formula for itself, with such suggestion as the plainly analogous biogenetic law may afford.

There are certain of the mores whose appearance in the young must be conditioned by their physical development. Speech, for example, is delayed until the organs of speech are ready to undertake the task; and the sex-mores lay but little hold upon the boy or girl until the period of puberty, with its subtle awakenings, is at hand. Illustrations of this kind exhibit anew the close and vital connection between the processes of organic and of societal evolution. But in general, and apart from cases as clear as these, the mores are taken on in a sort of order. A certain amount of originality may be seen in some children, but the permanent characters of the individual, or the permanent individual character, is the last development to be expected. That is, the child can be set down as belonging to a certain nationality, a certain class, a certain family, before its individual characters appear firmly settled. In general this is the order in nature, where in ontogenesis there emerge in sequence the characteristics of the class, order, family, genus, species, and so on. The analogy should not be pressed in detail, but there is evidently a course of development in the societal life corresponding in general trend to that in nature.

However, our chief concern is with the group rather than the individual; and if we may consider the immature as forming age-groups, their course of development can be seen to pass through phases roughly recalling those distinguished by scientists in the history of the race. It is often said that children are primitive — indeed, that they are little animals or savages; and the prosecution of the study of primitive peoples has led scientists to aver that savages are beings with the bodies of adults and the characters of children. The child of the adult phase of the race reproduces the adult of the childhood of the race. This is a significant principle, and it has borne fruit in application; for it has been seen that to understand primitive peoples, into whose life and circle of ideas we can set ourselves only incompletely, it is well to apply what we can find out from reminiscences and observation of the life and circle of ideas of the young of our own race. It is not intended in the present connection to try to catalogue exhaustively the correspondences between child life and

previous racial life; but certain illustrations
may serve to bring out the broad contention
before us, *viz.*, that there are features about
tradition which entitle it to take the place of
heredity in a formula of evolution, without
thereby disintegrating that formula as applied
to societal life.

The infant starts with no mores at all —
with no more than an animal could have.
"Natural" reactions upon environment lead
to the display of unbridled emotions, *e.g.*,
towering rages, which the mores forbid inasmuch as indulgence in them is prejudicial to
the societal interest. There is an utter disregard of *meum et tuum* which, in the absence
of discipline, that is, arbitrary enforcement of
the mores, persists for an indefinite period.
Later, the lying and thieving propensities which
no internal monitor can exist to check, since
they lead, apparently, to the satisfaction of the
individual interest, come into collision with the
rules of living in society; and they are then
done away with or modified from their baldest
form of expression. Among small children,
left to themselves, as among savages, force is
the only title to property; and well up toward
"years of discretion" the rule is that of the

strongest. Of religion there is none, or it is of the low order of excitement and fear. Taste is crude and undeveloped. Affection for children lends interest to the undeveloped stage; but much of the amusement and diversion derived from children lies in the incongruity of their actions and motives with those prescribed by the societal standards. It is possible to become fanciful or poetic in dealing with this topic — to say that part of the pleasure of the child in his primitiveness lies in his unconscious harmony with the life led during the childhood of the race; that certain hidden necessities of his being are supplied and then purged away through his submission to some elemental law of progression in growth; and so on. It might be said that, just as some important weapon of the past, like the bow and arrow, having lost its consequence, descends to be a plaything, thus social habitudes, now superseded, have come to cater to the tastes of childhood, in its play-world. But all this is at present imaginative construction, unverifiable, and it is not science. One can believe much or little of it in proportion to the development, in his make-up, of critical spirit as against imagination. But it is true that the mores are learned in some order, or at least some before others; and that, in general, the earliest learned are those earliest acquired by the race, and the latest learned those latest acquired. The societal life from birth to maturity (that is, up to the development of an independent individuality as a member of society) of the younger generation taken as a whole, is a rough and imperfect summary of the societal evolution of the race from the stage where the mores came into being on up to the present.

Peal says [1] of certain tribes in Assam that there is from childhood to marriage the most complete and recognized sex liberty; that morals begin with marriage, after which infidelity, he thinks, is rare. When the opportunity to marry presents itself, if they have the inclination, they pair off and settle down, going through some public marriage ceremony. "Thus the individuals epitomize their own race history. The 'marriage' comes as a restriction on complete sexual liberty." Many another survival in culture-history similarly epitomizes the past of the race.

The young within the primitive group will

[1] The "Morong" as a possible Relic of Pre-Marriage Communism, in *Jour. Anth. Inst. Gt. Brit. & Ireland*, XXII, 248 ff.

attain at last to the type of their elders' mores even if there is no deliberate education; what happens as societal evolution proceeds is, however, that there is an accumulation of mental reactions on environment, that is, a development of civilization; and it is through this accumulation that the status of the group rises. In order that it may be kept up, then, it is needful that the attention of the immature shall be drawn toward the mores regarded as most important for well-being, to the prosperity-policy of the group. But this is education: conscious and deliberate inculcation. One of the essential functions of education is to create appetites — wants. But wants are what begat the first of the mores,[1] and they are what impel the individual or the race toward self-realization. At first education consists in so disposing of the young that they shall have before them that which it is desirable to imitate: the boys are thrown with the men, the girls with the women, and they thus have held before them the respective types of sex-mores. The group in its natural division is the first schoolmaster. Later its function is partially usurped by the family, but in any group whatsoever the society

[1] P. 33 above.

itself always has a large voice in this matter.[1]
The home is at its best a sort of preparatory
school, where the penalties for infringement of
the mores can be tempered; but at length the
child outgrows this school, and it is found that
he will learn at the hands of his fellows what it
is next to impossible to inculcate otherwise.
The final test of adaptation to the societal type
and the heavy finishing strokes are given by the
real author and guarantor of the mores, the
society as a whole. But with the proper home
discipline and training, the young may escape
many of the chastisements of the cold-hearted
schoolmaster whom we currently refer to as
"The World" — by which we mean public
opinion measuring up the human product
against the mores.

In the most primitive form of education,
where life is far from complex, voluntary imitation is the predominant factor in tradition.
The mores are of a form approaching direct
reaction on environment and so are readily
tested. Many of them are easily proved to
be inexpedient through slight accidental or
careless deviation from their prescriptions;
the boy who neglects to observe the example of

[1] Cf. Todd, "The Primitive Family as an Educational Agency."

his elders when hunting is speedily corrected through physical hurt or hunger. To a very feeble intellect the advisability of silence, coolness, etc., during the chase appears unquestionable because plainly and concretely demonstrable. This is one of the reasons why the accounts of the education of savage children so astonish us; there are few blows or harsh words, but the young are docile, showing great respect for their models, the elders. Nature imposes the discipline and penalty, and so the young come to believe in their teachers and their precepts. They see the use of them. Inculcation results in adaptations whose expediency is self-evident. Further, the young do not dare question the authority of the mores, because of the traditional fear of their ghostly sanction. What makes the education of the modern child so hard is the distance and indirectness of attainment of recognizable satisfactions, and so the difficulty of correctly valuing the process; and likewise the absence of sanctions of an awe-inspiring order.[1] If a boy

[1] Cf. Lippert, "Kulturgeschichte der Menschheit," I, 226. He gives some examples. To illustrate the efficiency of the religious sanction of the mores, as resting on tradition and as endangered by education, the following quotation challenges reflection. "The proprietor of a village in the Konkan, while

had seen a companion lose his life because
he did not know a paradigm, or was sure he
would be damned if he could not demonstrate a
proposition, the status and repute of education
would forthwith rise in his estimation.   Fancy
the zeal with which astronomy would be studied
if astrological beliefs were current!

As long as the rising generation can proceed
at once to test the validity of the mores by
reference to the "exitus docet" principle —
whether the criteria of judgment are positive,
as in the case where bodily pain results, or
fictitious, though not less convincing, where
ghost-fear enters — so long will they accept

deprecating the action of Government in withholding support
from the higher education in the country, was on his own part
doing everything in his power to thwart the spread of primary
education among the peasantry of his village. On being asked
why he delighted in keeping the peasantry in ignorance, he
observed that it was a great blessing to him and other respectable
members of the village that the peasantry were left uneducated
and engrossed in superstition. 'At present,' he says, 'I am not
obliged to undergo the expense of keeping men to watch my
fields and my mango trees, because the village god does the duty
for me either without any remuneration or at the cost of a fowl
or a cocoanut.   But once the peasants are educated and made
to believe that the great Gramadev, of whom they are so much
now in awe, is nothing more than a block of stone capable of
doing neither good nor evil, thefts and impudence will increase
in the village, and there will be no end to my troubles.'"   Joshi,
"Household and Village Gods," in the *Journal of the Anthropological Society of Bombay*, II, 206.

the transmitted ways with all docility. But when a rational selection of the mores which they are to have inculcated in them has taken place through the antecedent action of minds more experienced, far-sighted, and subtle than their own, and they neither see the demonstration of resulting adaptation made in terms they cannot misunderstand nor yet have much respect for their instructors, natural docility is rare. Still worse if they know that educators are themselves in violent dispute over the value of what is taught; they cannot then, even though well intentioned, be more than half-hearted. There is none of this in the early stages, before much rational selection has come in, for the code is the code, and all adhere to it. Nor is there much of it nowadays in the matter of elementary education. The value of the "three R's" is seldom questioned even among those ignorant of them. Tests of a concrete order lie all about them. It is, as a matter of fact, unquestionable that a distrustful attitude toward "higher education" has some reason in it, especially in view of the assumption that a higher education is the right of all. Inculcation can be overdone; there is such a thing as being over-educated for one's walk in

life. By the use of coercion and under the
mirage of theories and ideals, lofty indeed, but
for all that not verifiable as to their results,
people can be led into the adoption of a code,
the possession of which is of advantage neither
to society nor to themselves. This is one of
the results of the entrance of the human mind
into the process of transmission in societal evolution.

The education of the young as a phase of
transmission of the mores is capable of extended
and intensive development. In fact, the program of education represents the greatest systematic attempt to put rational selection into
operation that the world has seen. It always
represents, moreover, the reflection of the existing social status, resting solidly upon the existing economic status, but played about by the
popular philosophy of the time. The root idea
of education cannot be anything else than to
raise up the rising generation to play its part
with the greatest satisfaction in its own self-
realization, in the societal environment and
within the societal code of its time. It is trained
to knowledge of the game by learning its rules.
It is put in the way of attaining what is thought
worth while according to the views held under

the existing mores. For the young in their irrationality there is performed an act of rational selection in the mores by which they are to profit.

When it is realized that the young have no extended fixed code, but are impressionable in a very high degree; that their interests are but accumulating and are not yet mature enough to guide them in their own selection; then the exceeding importance of making a good selection for them becomes apparent and, indeed, overpowering. It becomes a culpable thing to take the eye from the young and their destiny and fix all attention upon advancement of the profession of educator, or, still worse, upon petty personal vanity and private ends. The profession of educator demands self-abnegation; for its very essence, when it is followed truly, consists in straining all the powers to do discerningly and judiciously for those who are not yet ready to mark out their own destiny. Fortunately for the race's interests this profession has always harbored a number of devoted and unselfish souls.

Returning now to inculcation in the broadest sense, we may consider the matter of the education or "uplifting" of one class in a society by another (this being an intra-group process), or of one people or race by another (this implying inter-group relations). Both are cases of conscious and deliberate effort to transmit the mores. They are different in degree, but not in kind, and may be considered in conjunction. In educating the young, the idea, when it contains sense, is to equip them with the best code to enable them to fit the position in life which they presumably are to occupy in the future. There is here some room for hope that they may rise above the level of the standard of living of their birth-group; it is as desirable to incite in them all aspirations toward what they may attain as it is unwise to instill within them vague and restless yearnings for the impossible or highly improbable. Youth is elastic and adaptable, and it is perhaps just as well to set for it an aim slightly beyond the probable. When, however, it becomes a question of "elevating" classes and races, discretion and modesty of expectation must guide any safe policy.

Granted that the intent of the proposed "uplifters" is not disingenuous, what can they expect as results? To ask what they do expect would be to require a voluminous list of some of the chief chimæras and vagaries of history.

Perhaps a better question would be: How is it possible to get results at all? The answer comes directly out of our antecedent discussion about selection. The body of the mores tends to consist with the self-maintenance mores of the class or race in question. It thus reflects the status of the struggle for existence and for a standard of living, that is, the status of the relation between environment and the immediate reactions upon it.[1] Manifestly, if this is true, the point of attack should be either along the line of altering the environment (so that different reactions will be evoked) or in deftly hastening a lagging adaptation, chiefly of the maintenance-mores, to existent conditions. The former alternative represents a formidable task; the latter means that the uplifting agency shall formulate, for the proposed beneficiaries, a program of rational selection in their mores and then force or persuade them to adopt it. Here again great damage may be done by mistakes; and there is here more likelihood of error than in the education of our young. The latter are supposably being educated up into a set of mores, which, since they are our own, we may be assumed to know more about, both

[1] Pp. 141 ff. above.

as to their nature and the conditions with which they correspond; but when it comes to the other form of inculcation, we can possess no such intimate knowledge. This question of the possibility or advisability of trying to transmit our mores, as being the "best" — of trying to exercise rational selection for a group more or less alien to us — hinges largely upon an understanding of adaptation in the mores, and for the moment must be put over until that topic shall have been considered. In general, as has been seen, the effective place to take hold in attempting rational selection is at the bottom, with the mores whose adequacy is most quickly and convincingly tested, those of societal self-maintenance. Imitation will then ensue, for it is really due to the perception, mistaken or otherwise, of the pertinence of that which is imitated to the conditions of life.

Inculcation in all its forms rests ultimately upon imitation, which is the dominant means of transmission of the mores. A number of authors, whose interest lies largely in the psychological activities connected with imitation, have developed its "laws" in some detail.[1] This literature is readily accessible to those who

---

[1] Tarde, "Lois d'Imitation"; Ross, "Social Psychology."

wish to consider that aspect of the subject. Our interest in the present connection is in imitation as an agency of transmission rather than in an analysis of the mental state or action underlying it. Inculcation is really in good part an attempt to induce or force imitation; it is a deliberate attempt to pass something on to a prospective recipient. Having given this aspect of the topic before us some attention, let us now consider the wider and more general case of imitation where the initiative lies mainly with the recipient.

Imitation is discoverable well down in the organic world, and is conspicuous in the life of those animals which are nearest to man in the scale. The ape has given his name to a verb signifying to imitate. It is difficult to set up any line of distinction between the highest animal form of imitation and the lowest human form; here is yet another of those zones of transition, common along the path of evolution, and effective to blur boundary-lines between its grades. There is imitation in the concurrence as a result of which folkways form and acculturation begins; and no very great mental activity is necessary to secure such concurrence. "The ability to distinguish between pleasure and pain," says Sumner,[1] "is the only psychical power which is to be assumed." Mental reactions took place, and may seem in retrospect often to have been rationally directed; but the whole process was empirical. Mental operations were chaotic and unreliable; correctness of conclusion came, in certain categories of cases, from the fact that conclusive tests lay everywhere and unavoidable in the conditions of a rude life. But the validity of reasoning, as we view it, was not yet in the mental outfit any more than it is in that of children; and its absence made no difference in the matter of consensus. A more effective tool might be invented and then imitated to the real advantage of the imitators; but then, consensus in the irrational device, *e.g.*, the use of fire to exorcise ghosts, was just as ready; and some chance fortune or misfortune following action was quite as clear a demonstration as the observed superior efficiency of the tool could be. Convictions attained by the *post hoc ergo propter hoc* error were quite as unshaken as those arrived at by what would seem to us a flawless syllogism. That is, reasoning was scarcely yet even rudimentary; there was a general chaos

[1] "Folkways," § 1.

except where the irrefutable logic of facts forced a certain order. Whatever acculturation has been effected among primitive peoples has been received, except to some degree in the domain of the maintenance-mores, imitatively, irrationally, and almost unconsciously. Along the lines of concrete material interest, recognition of the superior values of others' mores is not seldom very keen, as it is in children; but elsewhere cases are vague and scattering among primitive peoples. The mores run out into ritual,[1] and the ritual, being supported by ethnocentric and religious sanction, becomes stereotyped and unchanging, so far as the will or intent of the society goes. The main lines of the mores once laid down by the rudest forms of societal selection,[2] variations in the mores susceptible of preservation and transmission are confined to ranges where, in the presence of constant tests, ritualization develops less readily.

Further, the primitive condition is one of group-isolation; and isolation is unfavorable to receptiveness and change. And so transmission goes on through the transfer of essentially the same set of mores from generation

---

[1] Cf. Sumner, "Folkways," §§ 67–68.   [2] P. 64 above.

to generation. This is perhaps the perfection of imitation as a machine-like agency keeping succeeding generations to the type of former ones. The very word "tradition" is synonymous with negation of change or "progress." In the absence of variation, competition, and selection, it tends, like heredity in organic evolution, to singleness of type and to monotony.

When, now, the society is compounded and differentiated in its elements, these elements are characterized by differing codes of mores. But some of the elements, or classes, seem to be succeeding in life — to be securing obvious material advantages to which others have not attained. Comparison of group-destiny, resulting in dissatisfaction, must have developed very early. Imitation of the ways of the successful group is the natural sequel. Thus the great and powerful — which means, in primitive times, the rich — might come to set the fashion in conduct, that is, their class mores might to some extent be transmitted down through the group. Vanity, fearing ridicule, powerfully strengthens this tendency. Individual variations in the mores of the great may come to an extension unwarranted by their intrinsic importance as judged in cold blood; and part of the mores of the upper classes of one generation or age may be infused into the mores of the lower classes in the next. Here is where, as we saw under the topic of rational selection, there is some chance, which can be utilized by the better educated part of the society, for alteration of the mores. But it should be kept in mind that the moving factor producing receptiveness to acculturation, and so allowing of the free action of imitation all through the body of the mores, is change effected in the maintenance-mores. Imitation as between classes is no more than superficial unless it rests upon a growing approximation of economic conditions and habitudes. However, there lie in the way of imitation of the economic mores fewer barriers of serious nature than elsewhere, and their imitation, like rational selection, may operate indirectly and has so done in history, with great effect.

Similar conclusions are reached when the case of inter-group or inter-racial transmission is considered. The imitation of a "higher" race's mores by a "lower," where it is not a superficial one, is generally confined to those of a more material order. Vanity may lead to the adoption of the clothing and some of the other externals of the more advanced by the less; the outer forms of religion may be taken over and accommodated to the essence of the old religious beliefs or superstitions.[1] But the history of missions has demonstrated that, with the development of sound reasoning by the side of, or in place of mere enthusiasm and good-will, the bare performance of the ritual — pattering prayers by rote as the Jesuits taught the Paraguay Indians to do [2] — is not deemed an adequate demonstration of uplift. The character of the inculcation has altered; so that medical missions (which aim to alter the conditions of life) and industrial missions (which aim to alter the maintenance-mores) have gradually come to prevail as against the older type. For reasons previously set forth, imitation is more easily secured within these ranges of influence; hence a wiser form of inculcation now aims at what is surest to effect acculturation in its best sense.

[1] "Experience shows," says W. R. Smith ("Religion of the Semites," pp. 355–356), "that primitive religious beliefs are practically indestructible, except by the destruction of the race in which they are ingrained."

[2] Watson, "Spanish and Portuguese South America during the Colonial Period," I, 272.

And if we confine ourselves to cases where inculcation is not attempted or thought of, the same general principle emerges. Imitation has, in a number of historic cases, taken place unconsciously and naturally along the line of the maintenance-mores, and has then been followed by an approximation of the "lower" to the "higher" mores all along the line. This is the normal process of acculturation, visible wherever real transmission of culture has taken place. The undeveloped race has had little difficulty in appreciating and taking over the maintenance-mores, the arts and crafts; here there is verification; things are seen to "work." And then the transmission of these has made inevitable the transfer of the rest.

For example, the Romans never tried to "assimilate" their provinces. All they cared about was the maintenance of their rule. But the provincials, observing the obvious superiorities of the Roman industrial and military organizations, imitated and fell in with these. They then began to see the advantage of the Roman language, Roman laws, and other of the secondary societal forms, and took them over naturally, almost unconsciously, and certainly

without compulsion or exhortation.[1] In general
the transmission of the mores which took its
course from Chaldæa and Egypt through the
Phœnicians, Greeks, Genoese, Venetians, and
others forms a grand illustration of the point
at issue.[2] None of these peoples had any mis-
sion to uplift western Europe; they were after
gain through trade. They operated exclusively
in the economic field, introducing first the products, then the processes of the superior arts
of the East. And with what result? With the
result of modifying at length the whole societal
structure of the West, by gradually transmitting
to it, as it developed a substructure capable
of supporting them, the mores of an advanced
civilization.

Trade has always been the greatest natural
carrier of civilization, and before the Age of the
Discoveries it operated in isolation from later
developed carriers to which, in the time since,

[1] This case of transmission is worked out briefly, chiefly after
F. de Coulanges, by L. de Saussure, "Psychologie de la Colonisa-
tion Française dans ses Rapports avec les Sociétés Indigènes,"
chapter on "Roman Gaul." Cf. James Bryce's two historical
studies: "The Ancient Roman Empire and the British Empire
in India" and "The Diffusion of Roman and English Law
throughout the World."

[2] This is worked out in some detail in Keller, "Colonization,"
Chap. II.

much of the influence exerted by commerce has been freely accredited. But it has remained the "handmaid of civilization," or, better, its forerunner. Trade and the development of communication, that is, the piercing of isolation, have always gone together. But it is precisely the contact of peoples ensuing upon the passing of isolation that allows of tolerance, mutual knowledge, and acculturation through imitation.[1] Commercial activities, however, aim directly at the creation and supply of material wants; this they do by suggestion through some form of advertising. They have enlisted imitation where missions, for example, have attempted inculcation; they have worked "with the grain," so to speak, where agencies of inculcation have gone against it. They have impinged immediately upon the maintenance-mores, where other agencies have attempted to begin with the secondary societal forms. So that here again we meet the now familiar case of effecting changes throughout the body of the mores by influence brought to bear upon the organization for self-maintenance. So

[1] This position has been worked out in an elementary form in Gregory, Keller, and Bishop, "Physical and Commercial Geography," Chap. XIII.

high an authority as Livingstone[1] suggested long ago that the way to eradicate slavery in Africa as he knew it was to better the transportation system. For the only agency for carrying burdens, capable of operating in the narrow paths and tunnels which penetrated the tangle of primeval vegetation, was the human pack-animal. "A single railroad," says Fabri,[2] who quotes the keen and clear-headed Scotchman, "is more effective to suppress slave-raids and the slave-trade than a dozen crusades, which would cost thousands of lives and would swallow up millions to just about no use." There was sense in these remarks, as the event is proving. The case shows the advantage of applying effort where it will count.

But it is plain that our cases are now verging over to illustrate adaptation quite as much as transmission. Imitation, I repeat, is really due in the large to perception, accurate or mistaken, vague or otherwise, of the pertinence of that which is imitated to the conditions of life. In particular is this the case with the imitation of the secondary societal forms — law, prop-

---

[1] Livingstone always had an eye to the civilizing effects of trade. See the last chapter of his "Travels in South Africa."

[2] Fabri, F., "Fünf Jahre deutscher Kolonialpolitik," p. 55.

erty, religion, etc., — for these do not possess vigor or effectiveness unless they are adapted to those primary forms that belong in the maintenance-mores. Fancy the maladaptation and so the ineffectiveness of the governmental system of the United States as existing among the aborigines of Fiji; it would be as superfluous in its complexity in those islands as the Fiji system would be meager and insufficient to discharge the corresponding regulative functions in this country. You cannot discuss any of these factors of evolution without having the end-result, adaptation, in the background all the time. Adaptation is the summing-up, as it were, of the activity of variation, selection, and transmission. Hence I have already infringed somewhat on the field of adaptation; and what I shall now have to say about it may be expected to throw back light upon the way already traversed.

CHAPTER VIII

ADAPTATION

SUMNER closes a manuscript essay on "Evolution and the Mores" as follows:

"In short, as we go upwards from the arts to the mores and from the mores to the philosophies and ethics, we leave behind us the arena on which natural selection produces progressive evolution out of the close competition of forms some of which are more fit to survive than others, and we come into an arena which has no boundaries and no effective competition. The conflicts are freer and freer and the results of the conflicts less and less decisive. The folkways seem to me like a great restless sea of clouds, in which the parts are forever rolling, changing, and jostling, as temperature, wind-currents, and electric discharges vary. We may confidently believe that there is not a cloud shape which does not correspond exactly to the play of forces which makes it, and we may be well convinced that no change of form takes place from time to time behind which there is not a change in conditions and forces, but we also know that the cloud shapes do not change in a series of any definable character and that they do not run forward in time towards some ultimate shape, but they change and change, rise and fall, ebb and flow, without any sequence and purpose. If they conform to the conditions and forces from moment to moment, that is the end of their existence. So it is with the folkways and the attendant philosophy and ethics. They conform to the interests which arise in the existing conjuncture, and that is all the sense they have."

In this quotation the presence of "progressive evolution" is denied in all the mores with the exception of those surrounding the industrial organization and so representing material control over nature. In speaking of progress Sumner usually began by admitting its presence in the arts of life and then went on to question its existence in the more derived societal forms. Because he did not find it there, he denied that there was evolution in the mores. Though he used to warn us that evolution included retrogression, and to illustrate pointedly by summoning us to witness that it was "the same force that made the stone go down and the balloon go up," he meant by evolution something more than adaptation.[1]

It would be a notable performance if we could prove that societal evolution is progressive, or that it is not progressive; but that is not the question we have set before us. All we have undertaken to study out, in the present in-

[1] See Note at end of this volume.

stance, is whether there is any evolution at all in the mores, irrespective of its general trend.

arwinian evolution is neither progress nor retrogression; it is both, as Sumner said. It is, in its final phase, adaptation to environment. But if we seek for adaptation in the folkways, we have Sumner's word for it that they are "subject to a strain of improvement towards better adaptation of means to ends, as long as the adaptation is so imperfect that pain is produced." [1] For our present purpose something like this is all we should care to prove. That there may be some "end" or "purpose" is, for the present at least, a matter of indifference; and as a philosophical concept it will remain so. It would seem, from what has gone before, that if improvement is admitted for the arts, processes, and systems of material culture, and if there is, as Sumner says, a "strain toward consistency," [2] in the mores, then there must inevitably result an improvement all along the line. This would be particularly evident if, as has been asserted above,

[1] "Folkways," § 5.

[2] There is, says Sumner ("Folkways," § 5), a strain in the mores "of consistency with each other, because they all answer their several purposes with less friction and antagonism when they coöperate and support each other." Examples follow.

the maintenance-mores are basic. Thus all
that is said about the secondary societal forms
consisting with the primary, since the latter
represent direct reactions on environment,
belongs to the topic of adaptation in the mores.

There is no course of reasoning, apart from
copious illustration, which can demonstrate
adaptation. And so our treatment, while it
cannot pretend to fulness, least of all to exhaustiveness, must be made up largely of cases
regarded as comprehensive and typical, it then
being left to the reader to confirm or reject,
from his own knowledge of other cases, the
principles derived from these.

But the conclusion to be reached at the end
may be set down at the outset. First, in general, if adaptation is admitted in the folkways,
their evolution on Darwinian lines is taken to be
thereby admitted. And if adaptation is seen
to be the result of the action of evolutionary
factors as treated in previous chapters, then the
Darwinian formula may be extended to cover
life in human society as well as in the organic
field. Second, and more particularly, the issue
of these chapters on adaptation is: Any settled
folkway (any institution, therefore, *a fortiori*)

is justifiable in the setting of its time, as an adaptation. It will be noted that the folkway is supposed to be a settled one, *i.e.*, a tried and preserved variation; a passing fashion or fad is a variation upon which selection has yet to pass, and neither in organic nor in societal evolution is there any reason to look for adaptation in variations prior to selection.

By way of a broad consideration bearing upon adaptation, to precede our illustrations, it may be said that the progress of knowledge has led to an unconscious revision of opinion concerning the justification of societal forms in their setting of time and place. In former ages of narrower horizons and small knowledge, when outlying races whom our predecessors were coming to know were seen to have mores diverse from those of the observers and often abhorrent to them, this was laid to perversity, depravity, or some other reprehensible quality. Naïve ethnocentrism exerted a sway so undisputed that, as a corollary to their own self-satisfaction, people believed that if the code of an alien society did not conform to their own, or if their own were not adopted at sight as self-evidently superior, it was because of wilfulness and even wickedness. Many survivals of this sort of belief yet appear; but on the whole such narrowness is now obsolescent. Ignorance, lack of opportunity, and lack of mental, moral, and social development were admitted, even by the Spanish crusaders in the New World, as excuses for non-receptiveness; and the tendency is growing to seek reasons for what used to be referred to stubbornness and caprice. The study of ethnography and sociology has contributed powerfully to this change of attitude. Such a racial experience and movement toward tolerance affords considerable evidence of a general nature looking to the establishment of our proposition about the justifiableness of the mores in their place and time — that is, toward the demonstration of adaptation in the mores.[1] For that growing

[1] Numerous cases could be assembled to demonstrate the damage done to primitive peoples by forcing or persuading them to give up their peculiar habitudes. For example, McGee (*American Anthropologist*, X, 338; Oct., 1897) shows that the attempt to repress the Indian custom of self-destitution through the *potlatch* wrought great injury to the natives. Ratzel ("History of Mankind," I, 284) describes the Hawaiian system of land-ownership, whereby the tribal rights of ownership went to the chief, the subjects cultivating for him or else giving him the first fruits of every harvest or rendering him compulsory service two days out of seven. He even received a quarter of all the wages earned by his subjects. The people were virtually serfs bound to the soil. The author comments: "A proof that

tolerance, which is greatest among the best informed, can be due to one thing only, *viz.*, an implicit, if not an explicit recognition of the fact that diverse or even offensive mores represent ways, more or less justified by their persistence over other ways, in which men have reacted upon environment in the struggle for life and for self-realization. Men have been in deadly earnest in this struggle and have done the best they could for themselves, dodging and twisting and turning to avoid the pain of non-adaptation; they have had no idea of being perverse, but have been preoccupied in the effort to secure greater satisfaction of tangible interests. I think it is safe to say that respect for the expediency of the mores of "lower" races increases with acquaintance, for they are not seldom remarkably apt to conditions; and where at first sight they look strange to the alien, they are often found, later, to have a great deal of sense, that is, to constitute felicitous adaptation.

this dependence was patriarchal, and not felt as oppressive, is furnished by the fact that the sudden abolition of it through Christianity has been indicated as one cause of the decrease of the population." A protest against the discontinuance of debt-slavery is recorded in Lewin's "Wild Races of Southeast India," pp. 85–86.

In this connection we have on record an instructive admission by Darwin himself. John Morley had accused Darwin of deficiency in historic spirit by reason of certain indignant expressions in the "Descent of Man" about slavery — expressions originating in quick sympathy and in "pathos" rather than justified by actual fact. Morley writes of Darwin:

"When, for instance, he speaks of the 'great sin of slavery' having been general among primitive nations, he forgets that though to hold a slave would be a sinful degradation to a European to-day, the practice of turning prisoners of war into slaves, instead of butchering them, was not a sin at all, but marked a decided improvement in human manners."

To which stricture the honest Darwin responds by letter at once:

"I believe your criticism is quite just about my deficient historic spirit, for I am aware of my ignorance in this line."

He then refers to the "Descent of Man" and continues:

"I felt that I was walking on a path unknown to me and full of pitfalls; but I had the advantage of previous discussions by able men." [1]

[1] Darwin, F., "More Letters of Charles Darwin," I, 326. The letter cited is of March 24, 1871.

If a man of Darwin's learning, discernment, and scientific spirit is forced to acknowledge such a piece of misjudgment of primitive mores, it is indeed a warning to lesser men against a glib and cocksure condemnation of whatever is alien to the spirit and practice of their age. Through several stretches in the "Descent of Man" Darwin was out of his field and was really not going behind the pronouncements of the philosophers to the facts. He slipped thus into the ethnocentric attitude and lost his "historic sense," that is, his perspective of evolution as a series of adaptations. He did not see that society's successive adaptations are as little the subject of moral judgments as is the adaptation shown in the thick skin of the pachyderm or in the "cruel" crunch of the tiger's jaws. Dispassionateness and an austere objectivity must characterize the attitude of any one who wishes to deal scientifically with society's evolution.

It is needed even more in the social sciences than in the natural because there is much more room for subjective bias in the former. Austerity and objectivity of attitude are all that can save the sociologist from merely painting yet another picture of his own code of mores under the guise of science — a performance in which he will be easily beaten by any second-rate newspaper-man, novelist, or preacher.

If there were no adaptation in the mores, there could scarcely be types of mores corresponding to certain broader and even to certain local conditions of the societies practicing them. But such types are to be found, and have been observed and distinguished from the most ancient times of which we have written record. These types naturally come out most sharply in the case of primitive peoples, whose reactions upon environment are most direct and simple. Let us begin, then, with them. We thus eliminate, for the time being, *sub specie experimenti,* those pervading intersocietal relations (of acculturation, etc.) which render the problem so complex and indistinct.

There is an active group of scientists in these days who have much to say about the influence upon the life of human societies of geographical environment. The fundamental and determining relations of a human group, these "anthropo-geographers" believe, are to its local natural surroundings: land forms, climate, flora and fauna, and so on. It is impossible to deny that these are fundamental conditions of life, for the human animal as for the rest of the organic world. This is particularly evident in the case of primitive peoples, for they are thrown into closer relations of dependence upon nature than are those whose civilization forms for them a certain protection against the action of nature forces.[1] Those who are unwilling to accept contentions about the influence of geographical environment upon modern civilized societies are often quite ready to concede the dominance of such influence in the case of the savage and barbaric.

Plainly the contentions of the anthropo-geographers root in Darwinism, for they reiterate and extend the idea of adaptation to environment. Cases of parallelism[2] constitute strong support for this position. But societal parallelisms can be genuine only where acculturation is absent, that is, in isolation. The constant tendency of civilization, however, is to break down isolation. This makes the appearance of clean-cut parallelisms under high civilization a matter of great rarity; indeed, the presumption is against parallelism, and this is reflected in the popular mind, whose tendency

[1] Pp. 67 above and 307 below.   [2] P. 216 above.

is always to assume acculturation. But this tendency means a reference of phenomena to reaction upon or response to the societal environment rather than the natural. That the type of environment has been altered by the increase and concentration of population and by the ensuing development of civilization is not infrequently lost sight of by the anthropo-geographer.

Civilization results also in the freeing of energy from the mere struggle for existence. But this freed energy — while it cannot exist and so can have no results, where the conditions of the environment impose a constant strain for mere livelihood; and while the form of its results must depend in last analysis upon what the environment can be made to yield for self-maintenance — may yet be employed in more or less whimsical or capricious ways, often, apparently, quite independently of any geographic "controls." Artistic creations, for example, while they may on occasion unmistakably reflect environmental influences, may on other occasions betray no connection with them. Or, again, this freed energy may be employed in reasoning about or drawing inferences from phases of the code which are themselves removed to some distance from concrete verification on the touchstone of actual life-conditions. Such reasoning might appear in connection with politics, ethics, or philosophy. The only reflection of the environment here shown would be its relatively favorable character, in that the struggle to live in it must be easy enough to admit of the release of this time and energy. Under a high civilization and a complex societal organization the multiplicity of causes assignable for a societal phenomenon — none of them verifiable through experimentation, while the ultimate test can be fended off except in the extremest cases — raises immediate distrust of anybody who asserts single or fundamental causes.

The only way to get at essential causes is to study societies whose life has not yet become so complicated; societies in such isolation that cases of parallelism stand forth clearly as such. It is the science of society which has shown us this way; and it becomes the conviction of the sociologist that what were fundamental causes once are always fundamental causes; that a catastrophic theory can be admitted here no more than in geology. Hence he cannot but believe that he may find, in the simpler life of a simpler society — there more clearly revealed where societal life is, as it were, in its lowest terms — the essential relations of adaptation in societal evolution. And it is hardly questioned that in such societies the adaptation is predominantly to physical environment.

I have gone to this length in contrasting conditions in civilized and uncivilized society partly because I do not want to ally myself unreservedly with the more ardent adherents of the sufficiency of geographical "controls." Among men the physical environment is not all; there is also the societal environment of fellow-men; and it is not the same thing as that of fellow-animals. And there is yet another environment which does not appear in nature at all: the imaginary environment of ghosts and spirits. The universal belief in the latter is one of the most significant cases of parallelism that we have. One of the greatest concerns in primitive life is adaptation to that environment. Fortunately for the non-complication of the case, however, the ghosts and spirits are entirely anthropomorphic, so that adaptation here is, in its general lines, like adaptation to the societal environment. No particularly divergent methods are called for; the differences are of degree and not of kind. But the whole case of adaptation, like that of selection and transmission, becomes more complicated from the very outset in societal than it is in organic evolution.

What I have found opening into view, then, in primitive cases of societal adaptation in comparative isolation is this: the organization for self-maintenance is of necessity directly adapted to natural conditions; then, on the principle of consistency in the mores, the rest of the societal system conforms to the maintenance-system. This latter contention has received a good deal of notice in foregoing pages, and, as I said at the time, will now receive considerable illustration here. It is evidently impossible, outside of an encyclopædic collection of facts that would take a life-time to assemble and set down, to make any exhaustive proof of my conviction. I shall limit myself, first, to one case of a primitive society's adaptation, hoping that that case, though extreme, will recall many others to the mind of the scholar or of the general reader. Then I mean to cite the adaptation shown by another type of isolated society, that of the frontier, which, in its rapid recapitulation of societal evolution, spans the chasm between the primitive and the civilized. I expect then to take up the type of adaptation peculiarly characteristic of a modern civilized society, and thus to conclude my essay. I do not by any means undertake to explain anything and everything in the societal organization by a formula; I am trying to indicate the main and characteristic lines upon which societal evolution operates, and no more.

The Eskimo are a primitive people occupying an exceptional habitat under exceptional conditions of isolation. They are likely to show the human form of adaptation in terms approaching the lowest and simplest. Something the same might be said of any group of desert-dwellers: the Australians, the Indians of southwestern United States, the Hyperboreans of Siberia, the Bushmen of Africa. Similarly isolated are mountain and plateau peoples, as in Tibet, or the inhabitants of inaccessible islands, as the Andamanese. Steppe and plains dwellers, as the Kirghiz, or the inhabitants of any other more open type of natural environment, being as a rule considerably less isolated, do not realize for us so good a "nature's experiment." And in what I have to say of the

Eskimo themselves I have reference rather to their former conditions, under extreme isolation, than to the present; for of late some of them, for instance in Alaska, have had their life-conditions altered, in this case by the introduction of the domesticated reindeer. The effect of such acculturation is analogous to the introduction of the horse among the Indians, elsewhere [1] referred to.

The pursuit of the struggle for existence, in the polar desert, must be carried on within the conditions offered.[2] Vegetable food can be had but in the smallest amounts, and there are no native condiments or intoxicants. Clothing must be of a single type as to materials, and the fact that it is worn for protection rather than ornament has led to the adoption of practically a single pattern. Shelter must be made out of the materials at hand; but these do not include any wood, at least for many of the groups, and

[1] P. 158 above.

[2] What is said here of the Eskimo is based chiefly upon the following ethnographical writings: Cranz, "Historie von Grönland"; Holm, "Ethnologisk Skizze af Angmagsalikerne"; Fries, "Grönland, dess Natur och Innevånare"; Nansen, "Eskimo Life"; Boas, "The Central Eskimo," in *Bu. Amer. Ethnol.*, VI; Nelson, "Bering Strait Eskimo," in *Bu. Amer. Ethnol.*, XVIII, pt. I; Murdoch, "Point Barrow Eskimo," in *Bu. Amer. Ethnol.*, IX. Also Ranke, "Der Mensch," II; Ratzel, "History of Mankind," II.

so the house-builder is thrown back upon animal products (bones and skins) and snow. If the Eskimo's methods of supplying these three main needs of man are scrutinized, there can be nothing but astonishment at the cleverness displayed in adaptation and at the perfection of the result. Civilized man cannot better them much, and finds it necessary, when sojourning in the arctic regions, to adopt the Eskimo ways. This is one of the best proofs of the success of their adaptation.

This same perfection of adaptation appears all through the maintenance-mores. Consider the effectiveness of the tools and weapons, sledges and canoes, of the hunting and fishing methods, and how this is attained by the use of materials whose working-up would seem hopeless for these purposes to peoples much farther advanced in civilization. Here we should note that what is to us a much superior adaptation may be a maladaptation on the Eskimo stage. The introduction of the rifle among these childlike savages has led, in some cases, to indiscriminate slaughter of game in the most wasteful way. The pleasure of firing off the "thunder-tubes" and of killing has been succeeded by scarcity, want, and misery.

One of the necessities of the environment is the conservation and generation of heat. To conserve animal heat the clothing and shelter of the Eskimo are wonderfully adapted. Similarly the heat in the bodies of slain animals is made use of by the drinking of the blood of those recently killed. And the lamp, simple as it is, and the methods of expressing oil without heat, represent apt reactions of the mind upon a situation the overcoming of which alone enables human beings to live in the environment.

Recurring for a moment to the matter of shelter, we find in the summer shift to the skin tent an adaptation which, when it was once partially given up at the instigation of aliens whose mores led them to think that people must be sedentary to be civilized, proved itself to have been entirely expedient.[1] Further, it is noteworthy that the use of snow for building purposes forced these people to adapt their architectural methods to the material, with the result that they developed the dome — a form unparalleled among primitive peoples, since it was not imposed in the life-conditions elsewhere. Similarly, it may be added, the environment forced adaptation to the snow in the form of

[1] Nansen, "Eskimo Life," p. 87.

snow-spectacles.   Each of these inventions and practices is in the highest degree rational as we look at it after the act, because it represents so expedient an adaptation.

But let us extend the field of observation somewhat.   In such an environment the relation of population to land is unique.   The former is thin and scattered (not over 35,000 all across America), and the groups are small and segregated.   The Law of Population [1] demands it.   This immediately alters the terms of the struggle for existence, for, in a sense, the social environment is removed.   Hence the mores corresponding to this situation of isolation at once appear; and we find a virtual absence of war (though legends, probably originating while this people occupied a habitat less peculiar, refer to war) and of trade.   Militancy on the one hand and commercialism on the other are nearly unrepresented.   In particular there have developed the mores of peace, so that Eskimo who visited Europe and saw the evidences of militarism thought of sending missionaries there to inculcate the loftier mores of amity.   The absence of trade caused the natives to remain strangers to its

[1] P. 24 above.

methods and they fell especially easy victims to the frontier trader.

Out of this elementary organization came few ideas and practices centering about the relation of mine and thine. Property in land other than in the use of a hunting ground could not enter the mores, of course, and the property that was recognized was largely communal where it was not, by the nature of the case, personal. Food was shared; sometimes in times of plenty with housemates alone, again in times of scarcity with all, even if they did not belong to the village. Among the Bering Strait Eskimo, says Nelson,[1] "if a man borrows from another and fails to return the article he is not held to account for it. This is done under the general feeling that if a person has enough property to enable him to lend some of it, he has more than he needs. The one who makes the loan under these circumstances does not even feel justified in asking a return of the article, and waits for it to be given back voluntarily." Here is an absence of recognition of capital, which operates as a brake upon the advance of the standard of living. Inheritance covers at the widest the tent and *oomiak*, the equivalent

[1] *Bu. Amer. Ethnol.*, XVIII, pt. I, 294.

of house and premises. Without going into further detail it is clear that the mores about property reflect the simplicity of the maintenance-mores, and respond, through them, to the conditions of the environment.

Closely connected with what has been described are the mores surrounding societal regulation. There is virtually no government, least of all control issuing out of militancy. The oldest man and the most skilful hunter enjoy prestige and the latter often holds the position of honor and danger, in that his *igloo* is the one farthest north. There is no slavery and so not even the beginning of classes. It will be recalled that enslavement comes out of war or debt, and even then does not become a settled institution till the agricultural stage is reached.[1] Here, then, where there is no war, where debts are of the sort indicated in the quotation just given, and where agriculture is out of the question, the very basis for the development of such a phase of the mores is absent. There is, of course, no law, but merely vague precedents. Crime is hardly defined, let alone a schedule of penalties. Government is, in a word, entirely elementary.

[1] P. 151 above.

The sex-mores harmonize with the conditions of societal self-maintenance. The "sense" of shame, which comes out of a realization of invidious distinction from the rest, is rudimentary. This goes back to the character of clothing and shelter as imposed by natural conditions. The custom — an adaptation whose abrogation has resulted in sickness — was to strip upon entering the *igloo*, the temperature of which, in the absence of ventilation, is high. Thus does the skin get a chance to exhale after being inclosed in fur garments whose pores have been filled up to exclude cold. But the crowded *igloo* allows of little or no privacy. Hence the naturalness of nakedness and the absence of shame; hence also a lack of chastity and decency as judged from the standpoint of codes formed under other conditions.

The preliminaries to marriage and the marriage ceremony, as practiced by most other peoples, are virtually absent. The sense of the ceremony is to secure publicity of the relation of a man and woman in matrimony; but there is no need here, in a small group, to perform a ceremony for that purpose. Capture survivals in connection with marriage indicate, however, that under other former conditions there was a more extended ceremony of marriage just as there was war and theft. The traditional form of marriage is monogamy, as is natural where poverty and a strenuous struggle for life exist; but, here as elsewhere, polygamy is practiced by those who can afford it. In fact, the situation exerts a stress toward polygamy: the death-rate of males, owing to the hazard of life, is very high, reaching, it is said, eleven per cent a year; hence the prevalence of mores calculated to foster the mating of all the potential mothers.

Endogamy is the rule and is inevitable under population-conditions; exogamy in any wide sense is impossible. However, there is a rather strong disapprobation of close consanguineous unions. Naturally enough there is no development of clans and tribes leading to confederation or wider societal organization. The position of woman, on this hunting stage, is not high; but the tasks that fall to her in the sex-division of labor are of sufficient significance in the struggle for existence to demonstrate her value and make her considerably more than a slave.

Children are few and their death-rate high; to find as many as three to a family is rather unusual. The nature of the food supply imposes protracted nursing upon the mothers, and so a long period of barrenness between successive births. But the few children are valued highly and receive affectionate treatment. They are untrained except by the exigencies of life,[1] but by these are disciplined into a characteristic mood of patient endurance, courage, and good nature. It is clear that from all aspects the sex-mores and family life are elementary in form and that they reflect the organization for maintenance at every turn.

The severity of the struggle is such that but little energy can be freed for the arts of pleasure. What there are are very simple and childlike. Such also is the religion of the region. The late George Borup used to say that the Eskimo did not believe in a God, but that they did most fervently believe in a Devil — they had nothing to expect from the former and everything from the latter. Burial customs, and other of the mores connected with the dead, witness to the presence of simple animistic beliefs, but these are very crude and vague. The spirits of the ice, the *aurora borealis*, etc., are feared; and explorers have always had great difficulty in

[1] Cf. p. 228 above.

inducing an Eskimo to go far north. The whole of the Eskimo religion betrays the over-powering fear of ill — a fear entirely justified amidst the life-conditions. What little cult there is is of the negative sort: avoidance and exorcism. The strain of the struggle for exist-ence and the prevalent poverty admit of no overplus of time or wealth to be spent on the cult. Thus the religion too reflects indirectly the nature of the struggle in the local environ-ment.

Here then is a primitive and isolated society a survey of whose mores indicates beyond ques-tion the close adaptation of the maintenance-mores to environmental conditions and the consistency of the other mores with those of self-maintenance. Even in an almost unique case of primitive isolation like this it is possible to discover many a detail of the mores that shows no adaptation to or consistency with anything else in particular. Such are the variations in the mores that are either of too little significance to living to have incurred selection, or are too newly arisen to have yet come under it. In dealing with adaptation it is necessary to satisfy one's self with respect to the larger and more massive groups of mores and institutions. It must be recalled that in the social sciences we cannot experiment; that the experiments nature performs for us are generally over a long perspective and little illuminative of detail; and that the most we can hope for, on our present stage of knowledge, is a high degree of probability as to the working of the larger factors in the life of society.

CHAPTER IX

ADAPTATION (*Continued*)

IF what has just been said of the study of a primitive and highly isolated society is true, it is even more pertinent to bear it in mind when we approach the question of adaptation in a more complex, less primitive, and less isolated one. Such a society is the frontier group or the colony. It is a sort of primitive civilized society, and yet of course it is not really primitive. As a laconic student once put it: "Frontier society is not primitive — no more than a man who puts on overalls and fixes his plumbing is a common laborer. The frontier puts on the overalls." The Greeks used to call the city which sent out a colony the metropolis; they made a good deal out of the analogy between the colony-metropolis relation and that of child and mother. We too speak of the mother-country, and also refer to the colonies as daughters. There is usually a good deal of truth in popular and proverbial sayings — they often embody tested folk-wisdom — and the present instance has something in it that is deeper than figurative. The frontier society is a child-society, with capacities for rapid growth that may enable it to equal the mother-society in point of size, strength, and culture within a relatively short time. Here is, in actuality, a case of recapitulation, where the evolution of the child-society presents a summary, abbreviated and with a number of omissions, of course, but in the main, and in the order of rehearsal, faithful to the course of evolution traversed by the adult society. This is the reason why the study of frontier societies is so important to the science of society; it, together with the study of savage and barbaric peoples, provides actual evidence for societal evolution and constitutes a point of detachment and perspective from which to view that of which we are all a part.

Says Sumner [1]:

"When we gather together the observations we have made, showing the advance of the entire social organization from the colonial settlement up to the present time, in all its branches — the industrial system, the relations of classes, the land system, the civil organization, and the

[1] "The Challenge of Facts and Other Essays," p. 331.

organization of political institutions and liberty — we see that it has been a life-process, a growth-process, which our society had to go through just as inevitably as an infant after birth must go on to the stages of growth and experience which belong to all human beings as such.   This evolution in our case has not been homogeneous.   The constant extension or settlement into the open territory to the west has kept us in connection with forms of society representing the stages through which the older parts of the country have already passed.   We could find to-day vast tracts of territory in which society is on the stage of organization which existed on the Atlantic coast in the seventeenth century; and between those places and the densest centers of population in the East we could find represented every intervening stage through which our society has passed in two hundred years.   This combination of heterogeneous stages of social and political organization in one state is a delicate experiment; they are sure to contend for the mastery in it, and that strife threatens disruption.   As I believe that this view has rarely received any attention, it is one of the chief points I have wished to make in surveying the advance of social and political organization in this country."

From the evolutionary point of view a colony is a reversion.   But that means no more than that it is an adaptation to a set of conditions out of whose range older societies have passed.   Reversion is as much adaptation as is progression; both are evolution — "it is the same force that makes the stone go down and the balloon go up." What produces the reversion is the change from a partly artificialized environment to a natural one. The frontier society then adapts itself to the crudity of nature sacrificing, as we shall see, much of the civilization which it had, in favor of forms of adaptation which (and this, as I have said, constitutes a remarkable case of parallelism) are successful as they resemble those of the natives Acculturation takes place, strange to say from the "lower" toward the "higher" race thus the colonists in New England, as we shall see, "Indianized." In short, the colony or frontier society presents cases of adaptation as striking in their way as those of a primitive society of savages.

There are two major types of frontier society which are differentiated one from the other upon the basis of the character of their responses to physical environment, chiefly climate. These are what I call the farm-colony (of the temperate zone) and the plantation-colony (of the tropics).[1] This differentiation represents a fine underlying case of adaptation, and I shall return to it. The two types have generic likenesses in characters common to all frontier

[1] Keller, "Colonization," Chap. I.

societies, and these I wish to bring out first of all. This can be done most readily by first throwing the temperate zone colony into contrast with the mother-country, and then the plantation-colony into contrast with the farm-colony. Hence the bulk of description will fall to the latter form, which I have chiefly in mind in dealing with the generic characters above referred to. The temperate colony is, in a sense, a simpler case, because it has more conditions in common with the older society; there is, in general, an identity of climatic environment and of race between them.

The significant condition in any country is the relation of population to means of subsistence, that is, in general, the relation of population to land. The Law of Population for men, as we have seen,[1] is that population tends to increase up to the supporting power of the environment (land) on a given stage of the arts and for a given standard of living. If, now, there is a great deal more land, the arts may decline or retrograde, especially if the standard of living also declines, and yet the population may increase rapidly. If this were not so, new lands could scarcely be settled. In actuality

[1] P. 24 above.

the arts will not decline to the level occupied by them among the adapted natives of the new land; some of the arts brought in from the older countries will be applicable at once to the new conditions and will constitute a support for population. The conjuncture is overwhelmingly in favor of men as against land.[1] Hence a temperate colony shows, as a rule, a rapid increase of population. The struggle for self-maintenance is relatively easy.

However, with the retrogression of the arts and of civilization there is a lessening of protection from natural selection.[2] The frontier is no place for the weak; in fact, there is a sort of preliminary selection before emigration, as is shown by the overplus of men and the relative absence of women on the frontier. The weak and timid are repelled by potential danger and suffering. Counterselection is much less frequent under such conditions. A temperate colony generally shows a high physical quality of population — youth, strength, and vitality — as well as a rapid increase of numbers.

The standard of living cannot but decline; indeed there exists no possibility of realizing

[1] Sumner, "The Challenge of Facts and Other Essays," pp. 111 ff., 293 ff. [2] Pp. 66 ff. above.

a high standard. Beyond physical comfort there is little to strive for. But a lowered standard of living means at the least less of voluntary restriction of numbers. Marriages are early and families large. Often a low birth-rate is transformed into a high one upon removal to the frontier and upon the consequent cessation of the struggle to realize a standard of living no longer represented in the environment.

These are some of the major relations between population and land as they work out on the frontier where land is plenty and men scarce. The unoccupied land constitutes a sort of vacuum into which population tends to flow, not only by immigration, but also by natural increase. This capacity for rapid growth insures that the colony will presently catch up with the older countries, passing swiftly through stages through which the latter have moved slowly, and thus recapitulating, so far as numbers and the consequences of numbers go, the evolution of the older groups.

The general case may profit in clearness by a parallel illustration where the conjuncture of men and land was suddenly and violently altered by entirely different forces. The Black Death of 1348 acted through all Europe as a reducer of population, and during the ensuing century the effect is visible on the whole societal organization in all its institutions and customs. Similarly in the case of the Hundred Years' War in France. Great areas returned to wilderness and then were re-occupied as free land, that is, by immigration. Those who went out to reclaim the waste or into military frontier stations were freed from traditional tenure. There arose a new contract-relation which was rational and advantageous, that is, adapted to the altered conditions of the struggle for existence. All through the last centuries, as the result of the discovery of new lands, these phenomena persisted. Steam navigation really added millions of square miles to western Europe, and the effects of the ensuing underpopulation upon the societal structure have been characteristic.

Recurring to the population conditions of the temperate colony, we find that the men are scattered in segregated small groups over wide areas. This results in an isolation and a diversity of interests that set the groups off one from another in somewhat the manner of primitive tribes and communities. Socializing forces are weakened and there appears a sort of lapse into the atomistic form. The subdivision of the population into small groups makes it seem at times as if the patriarchal family had come again into its own. Then with the filling up of the country these small groups are obliged to coalesce under pressure, and so again there appears a recapitulation of the process of compounding societies which has produced the integrated state form.

On the basis of these population conditions now surveyed, and the maintenance conditions resulting from them, there has grown up a characteristic set of mores of the frontier type. As we come to particularize somewhat more, it is convenient to have some definite frontier society of temperate climate in mind. There are a number to select from; but a more comprehensive picture can be gained if we take a European colony rather than one founded by another race; and a modern colony rather than an ancient. I have in mind chiefly the northern English colonies in what is now the United States, though I do not intend to exclude the offshoots of these colonies toward the American West.[1] In general it is the colonization of this

---

[1] I cannot say just where the information upon which the following is based came from. Most of the facts are well known,

country, with especial reference to the northern members of the Thirteen Colonies, that I have in mind in what is immediately to come, about frontier societies and their conditions and habitudes. Here it is possible to see an extraordinary case of adaptation in the mores.

In the American colonies of the seventeenth and eighteenth centuries the industrial organization reflects with fidelity the condition of the struggle for existence and the competition of life. The commanding phenomenon is the simplification of the maintenance-mores all along the line; they came to resemble those of the adapted natives. There was a drop from the manufacturing or commercial economy to the agricultural, to the pastoral, and even to the hunting stage. The characteristic develop-

having been set forth in such books as Bruce's "Virginia" and Weeden's "New England," as well as in more extended histories of the period. A great deal of what follows goes back to Professor Sumner's courses in "American History," parts of which have gotten into print in his collected essays ("War and Other Essays," "Earth Hunger and Other Essays," "The Challenge of Facts and Other Essays" — especially the essay on "Advancing Social and Political Organization in the United States"), but the most of which is preserved only in the notebooks or memories of those of us who were fortunate enough to study with him. The conclusions about tropical colonies have been derived from reading represented in good part in the bibliography of my book on "Colonization."

ments of an advanced industrial organization
— specialization of function, with consequent
coöperation — fell back to much more rudimentary forms. And with the return of the
organization of labor to less complex and evolved
forms went that of capital; this prime necessity of an advanced civilization existed in small
amounts and was largely unavailable under
the ruder and less artificialized life-conditions.

For our purposes the organization for societal
self-maintenance may be readily viewed under
the captions of production, consumption, and
distribution. The first of these was the result
of a direct and largely unconcerted assault on
the flora and fauna. Nearly everybody was a
farmer or cattle-raiser, whatever his ostensible
occupation; the absence of the narrow specialist is as striking as the presence of the jack-of-
all-trades. Of manufacturing there was little
except along the line of home-industry for
local use; and this condition was encouraged
or even enforced by the mother-country, which
was bent on a division of function as between
itself and the colonies. This division of economic function was in the nature of things,
especially in the early life of the colonies, and
clearly betrays the colonial society in adaptation.

Consumption was of simple things locally produced. The food was grown for the most part by the eater of it, clothes were home-spun where they were not made of the pelts of animals, houses were the product of the personal industry of their future owners, aided by coöpcration in the "raising" process on the part of neighbors. In a word, the consumption of materials for self-maintenance took place near the spot where they were produced.

This was necessarily the case in view of the rudimentary character of distribution. Roads and vehicles were alike primitive, and the waterways, though these were not well-charted and safeguarded, provided the best means of communication. In view of these conditions exchange took on of necessity a primitive type. It was developed on a small and local scale, and was largely barter-trade. The standard of value became the characteristic local commodity (beaver-skins, tobacco) and the circulating medium (wampum) was adopted from the Indians. Peddling introduced a certain amount of goods from outside, and that stock transitional form to the market, the local fair, became characteristic of these as of all frontier societies. Frontier trade, lacking the protection of the market, whereby prices are standardized for all, and relying on defectiveness of information, showed its usual phenomena of haggling and overreaching.

The conjuncture, or relation of supply and demand, as between land and men being in favor of men, rents were low and wages high. Land could have little rent where any one could become an owner with comparatively little effort; wages had to be high, for the laborer must be paid for renouncing independent ownership and cultivation ("ophelimity"). Again, in the relative absence of capital and under the risks attendant to lending, the rate of interest was high. All these conditions represent reactions on environment of an adaptive order, and afford parallelisms to the adaptations of primitive societies.

In short the struggle for existence was cruder and coarser; but with the forces at hand there was presently an exuberance of simple satisfactions. Quantity came to be prized over a quality that could not be attained as yet; the standard of living was a simple and unrefined one. Hence the criticisms of cultured travellers in the colonies and even, later, in the United States; hence also the long subsequent strictures of Dickens, Spencer, Brunetière, on American ideals.

Part of the self-maintenance organization in the colonies was military, and this also shows a primitive type. Methods of warfare were largely those of the Indians, and that they were better adapted to the local environment than those of the English army, for example, was shown in many an engagement, small and large. The feeling about chivalry in conflict and for human life seems to have suffered a change. Certainly a reversion to Indian mores took place from time to time along these lines; indeed, it was in respect to methods of warfare practiced by the colonists that the word "Indianize" was used. Various forms of barbaric cruelty have been exhibited by settlers in America, and even so savage a practice as that of scalping. Some of these cases plainly come under acculturation through imitation, but along the main lines of warfare the illustration of adaptation in the mores is clear.

Self-maintenance being assured, a society turns naturally to self-gratification — to the satisfaction of wants over and above subsistence. The satisfaction of these, on the frontier, is of a crude and "boorish" order. High-class drama or opera are out of the question and are not desired; art in its more refined manifestations would be entirely out of place. The type of amusement is rather the itinerant "show" and, more locally, the homely festivities accompanying corn-shucking, house-warming, horse-racing, turkey-shooting, and so on. Gambling has been a constant phenomenon of frontier life, as of that of the savage. The satisfaction of curiosity has been attained by the practice of a wide hospitality, and by an extensive development of gossip still characteristic of isolated communities. Litigation has offered a wonderful field for self-gratification; and the baiting of the English governor was also a favorite pastime in the colonies, ranking with bear-baiting. Politics formed a most promising field of diversion. Even the practice of religion became a means of self-gratification; church-going was one of the few means of gratifying the so-called "social instinct" in an environment of comparative isolation.

The mores of self-perpetuation exhibit the same phenomena of adaptation. Woman's economic function was arduous, but honorable;

hence her position was high. On the frontier women are scarce and valuable.[1] There was less of property-consideration in marriage than in a more developed community and the wife was more of a help-mate. Children, under the existing conditions of easy support, brief education, scarcity of labor, and the like, were economic assets, where, in a more developed society, they are liabilities. Families held together in the patriarchal style; much was made of the extended relationship created by a necessary endogamy. The maid-servant, often the daughter of a neighbor, was "help" rather than hireling, and belonged to the *familia*. One of the curious details in the sex-mores, of which a number could be collected, is afforded by the practice of "bundling." [2]

The religious code was not a gracious one, but suited, rather, to the ruder conditions of life. Like that of the Boers, another colonial society showing characteristic adaptations, it

[1] This was chiefly under the agricultural economy. In early Kentucky "the female sex, though certainly an object of much more feeling and regard than among the Indians, was doomed to endure much hardship and to occupy an inferior rank in society to her male partner; in fine our frontier people were much allied to their contemporaries of the forest in many things more than in their complexion." Butler, "Kentucky," p. 135.

[2] Cf. Sumner, "Folkways," §§ 576–581.

was of the Old Testament type, with certain
ingredients of superstition and persecution, as
in the witch-craft beliefs.   But a common creed
formed a socializing element where there were
few such, and in the face of a harsh struggle
against nature, natives, and external enemies,
conformity was insisted upon.   There was no
refinement of religion; ingenuity and reflection
went for the most part into the affairs of practical life.   The religion of the colonists seems
never to have interfered much with the realization of less spiritual purposes.

It has been noted above that the settlements
produced locally for local consumption.   There
was in them an economic self-sufficiency and
independence that worked out into a general
mood corresponding to these qualities.   This
mood was easily exaggerated into testiness and
irritability.   The colonists were quick to suspect an infringement upon their rights and
liberties and prompt to resent it.   They were
hard to manage.   This was due in part to their
life of economic independence, which contrasted
strongly with tropical colonial life, as we shall
soon see, and partly also to other factors, to
which we now turn, which developed as adaptations to environment.

I have quoted Sumner elsewhere [1] on the topic of colonial democracy; the following passage enforces much that is brought out in the other citations, but from a somewhat more general point of view.

"All men are easily equal when all are substantially well off, because the social pressure is slight; it is intense social pressure which draws the society out into ranks and classes. The relaxation of social pressure lets the ranks and classes come together again.

"The three classes which form the skeleton of any aristocratic system, that is, of a system in which classes are widely separated from each other, are landlords, tenants, and laborers. The landlords are the holders of the land. The tenants are the holders of capital, because the land must be intensively cultivated, which cannot be done without capital. The laborers are those who have neither capital nor land, and who seek a livelihood by putting personal services into the industrial organization.

"If the population is dense and the land is all occupied, the possession of it is the possession of a natural monopoly of a thing which is in high demand. The landowners, therefore, possess an immense social advantage. The tenants and the whole middle capitalist class, which stands on the same social plane with them, possess the second social advantage. The laborers are those who possess neither. The three, therefore, are widely separated one from the other as respects the conditions of material well-being and earthly happiness.

"Suppose then that new social power is won — let it be assumed that some new mechanical force is obtained or that new areas of land are made accessible — what is the effect on the position of classes and on the relative difference in the status of classes? Plainly the social pressure is relaxed. The landlord finds that his monopoly is no longer worth as much as before, because the

[1] Pp. 147–149 above.

supply of it has been greatly increased. His rents decline, and his tenants refuse any longer to be tenants because it is so easy to obtain land and become their own landlords. In their turn they find it harder to hire laborers; for when land is abundant intensive cultivation is no longer necessary and no longer pays. Capital is no longer indispensable for the cultivation, or a small amount of it will suffice. The laborer, therefore, is no longer differentiated from the other classes. He can easily obtain land and also the minimum of capital necessary to cultivate it. Thus the landlord comes down to be his own tenant and his own laborer. The tenant owns his own land and is his own laborer. The laborer becomes his own landlord and his own employer. The three classes have melted into one. It is no longer worth while to own a large estate in land, for the owner could not economically exploit it. A substantial equality of all on the middle rank is the inevitable social consequence, with democracy and all the other cognate political results.

"At the same time, since capital is no longer so necessary to cultivate the ground, and since the accumulation of capital goes on with constantly greater rapidity, on account of the large proportion of the product to the labor under the new state of social power, and since the capital cannot be made productive without new supplies of labor, the men are on all accounts in demand and are worth more and more when measured in capital. The class, therefore, which was, under the first supposition, the worst off, obtains under the second supposition the command of the situation.

"Is not this the correct interpretation of what we see going on about us? If it is, then the dogmatic or philosophical theorems, instead of being the cause of our social arrangements, are only the metaphysical dress which we have amused ourselves by imagining upon them. We are not free and equal because Jefferson put it into the Declaration of Independence that we were born so; but Jefferson could put it into the Declaration of Independence that all men are born free and equal because the economic relations existing in America made the members of society to all intents and purposes free and equal. It makes some difference to him who desires to attain to a correct social philosophy which of these ways of looking at the matter is true to the facts." [1]

This democratic form of the colonial mores arose, then, in the nature of things, and was not the product of the happy thoughts or superior wisdom of the fathers. If we look to the nature of immigration into the colonies, we see that there was no place for the wealthy, nor yet for the confirmed pauper; neither for the aristocrat nor for the retainer. Both of the extremes of society were repelled from an environment where wealth could not profitably be invested in large amounts, where control over others, owing to the ease of evasion, was weak, and where every one had to take care of himself without much aid. Hence the bulk of the stream of immigration was composed of the sturdy middle class, or, in the case of the redemptioners, for example, of those who could speedily rise to that class. The history of the grants to aristocrats shows how out of keeping were aristocratic institutions in the new land. Again, though it was tried, slavery could not flourish long in competition with free labor;

[1] Sumner, "The Challenge of Facts and Other Essays," pp. 156–158.

and slaves were few and not sharply distinguished from the rest of the population. Few humanitarian devices could be developed or supported under the conditions of small and detached towns, and public dependents were therefore relatively unrepresented.

All these conditions made for homogeneity of population; and the situation as respects rents and wages, added to these, produced a general feeling of equality — of being as good a man as the next one — that was reflected in the mores throughout the whole societal system. Hence a democracy inherent in the nature of things and representing the type of adaptation of society to environment inevitable for success in the life-conditions. Hence the type of political system: town-meeting, selectmen, and the other familiar forms. Hence also the impatience of control, revolution, and the founding of a democratic republic. I shall return to this point as one of comparison between a colony of the temperate zone and one in the tropics.

As in the case of the rest of the mores, those which constitute regulation took on a simplified form. Government was local and its machinery simple. The democratic tendency led to the

lodging of temporary authority in the hands
of selected men (selectmen); but this did not
prevent the strong man, who was able to direct
and lead the assault upon a harsh environment.
from exercising great influence. As in time of
war, the military chief got political position.
The *posse comitatus* was in evidence. But, in
general, law and regulation were relaxed except
in the matter of essentials, and there they were
strengthened and made more arbitrary. There
were not so many petty regulations on conduct,
for those appear only when men are crowded
together and societal adaptation calls for a
nicer delimitation of spheres of rights. Regulation was, as it were, less extensive and more
intensive; including less, but operating with
greater severity in a smaller field.

It is not necessary to go into full detail in
connection with the adaptation of the mores of
a frontier society to its environment. It is
the point of view only which I am trying to
bring out. I think that it emerges with especial distinctness in the case of frontier justice
and with the development of that topic I shall
end this sketch of the mores of the frontier
society in the temperate zone.

We are familiar with the beginnings of justice in the individual settlement of private wrongs. This is a primitive form; very early in the evolution of societal institutions the ruling power takes over such settlement. But on the frontier there is a return or reversion to the primitive form; and it is clearly a case of adaptation. Even where the matter is not purely individual, there exists the hereditary feud, or a summary settlement, as in "lynch-law," or something approaching court-martial.[1] This is not always due to the absence of law or of a recognized government; it is generally referable, where it is not merely an outbreak of unregulated passion, to a dim recognition of the maladaptability of the machinery of justice as developed in a more artificialized environment. The law's complications, expenses, and delays detract from its usefulness anywhere, but they render it entirely ineffective where the society has no apparatus of police and prisons, and where the business of living presses too heavily upon the population to admit of long-drawn-out processes. Hence frontier justice is summary, and approaches the primitive type.[2]

[1] Cutler, "Lynch Law"; cf. Langford, N. P., "Vigilante Days and Ways"; Haydon, A. L., "The Trooper Police of Australia."

[2] A graphic study in frontier justice is worked out in Wister's novel, "The Virginian."

It appears sometimes that the severity of penalties exacted on the frontier is all out of proportion to the magnitude of the offense; horse-stealing, for instance, has often been punished, on the frontier, by death. But this case demonstrates the justification of a social practice — to a city-dweller as unjustifiable as many a habitude of savages — in its setting. Horse-thieves were at least potential murderers, because they took from a man what was all but essential and often quite essential to life. And so the retaliation was proportioned to the peril rather than to the actual result.

In connection with crime and punishment the case of the so-called convict or penal colony affords a striking case of adaptation. One of the surprising facts about such colonies is that presently they become respectable and, as in the case of Australia, develop into seats of culture. It is also found to be true that the deported criminal often became a useful member of the colonial society. Some criminal colonies have presently served notice upon the deporting country that they will receive no more convicts. But the explanation of this situation is, from the evolutionary standpoint, simple enough. A criminal is one who departs in his code from certain essentials of the mores, conceived so to be by the society in which he lives. But often what would be a crime in a settled, civilized community would not be even a misdemeanor on the frontier, with its primitive code. In early Greece Herakles was a god; but an individual who duplicated his boisterous practices in an environment of civilization would be imprisoned. The rude strength which cannot brook the restraint of the code developed in an artificialized environment may be an asset in direct contact with the rudeness of nature. If crime is a sort of atavism, then the criminal may be well adapted to the code of a less evolved stage of societal evolution. The stock gets a new start and is found at length to have adjusted itself to evolved stages of the code; it thus ceases to be criminal. And if, as was so often the case in the past, a person was branded as a criminal and deported for slight infractions of the arbitrary code of a ruling class (political criminals), the explanation becomes still easier; in such cases the line between the traitor and the patriot can scarcely be drawn apart from a knowledge of who has gained the power.

It would appear, then, that the frontier

society is obliged to adapt itself to a dis-artificialized environment and that in so doing it presents a case of retrogression as seen from the viewpoint of "civilization and progress." It would seem sometimes that the "new" society has to learn all the lessons of civilization over again; that it cannot but fall into the administrative, financial, and other blunders out of whose heritage of ill the more developed societies have painfully extricated themselves. But thus does the child, in his ontogenesis, repeat the experience of the race. Experience of these ills is really essential, for only by reason of the pain of maladaptation is a better adjustment to constantly changing conditions attained.

I wish now to make a brief contrast of the mores of two types of frontier society, whose basic distinction is that one develops in the temperate zone and the other in the tropics. It is consequent upon adaptation to diverse sets of environmental conditions, as I take it, that these differences in societal forms ensue; and if this is true, then we have another clean-cut case of adaptation in the mores.

[1] This contrast is worked out in greater detail in the author's "Colonization," Chap. I.

When Europeans migrate to another part of their own climatic zone, the comparatively slight change is a beneficial one, other things being equal, for the human organism. Further, the competition with the natives is short and the coercion to healthful activity is beneficial. But it is quite otherwise when migration is to the hot lands: the organism feels a violent change; the menace of the acclimatized natives is always present, inasmuch as the white man cannot sweep them aside in competition; and there descends on the powers of the immigrant a lethargy which renders him economically of small use.

These circumstances are reflected in the conditions of population. In the temperate zone the immigration is comparatively copious, and, at least after a short time, includes both sexes. Natural increase is rapid and there are few half-breeds. The population is biologically homogeneous, and it is vigorous and healthy. In the tropics, on the contrary, immigration is slight, and is almost exclusively of males. Natural increase is also, for the immigrant race, inconsiderable, but mixed matings result in many hybrids. The immigrants cannot work the soil and are neither healthy nor vig-

orous. The population is biologically heterogeneous.

In the organization of the struggle for existence the contrast is sharp. In the temperate
colony the products are essentially those of
the home land. They constitute a variety of
necessities. They are raised on the small scale
on small freeholds or farms, and these are locally
self-maintaining. Labor is in the hands of the
immigrants and it is vigorous and carefully
expended with a view to preserving the soil.
There is little danger of scarcity because of the
variety of crops raised, and there is developed
a consequent economic independence of the
mother-country and, indeed, of the rest of the
world. In the warmer countries, on the other
hand, the production is of luxuries which are
few and costly, and which must be raised on
the grand scale to be profitable. But the white
man cannot do the work and the acclimatized
native does not want to do it. Hence the
"native labor question," which has generally
meant enslavement in some form. This mode
of organization has led to uneconomic methods
of treating the soil; the white men are only by
rare exception more than temporary sojourners
in the land, and therefore do not naturally

develop far-sighted policies in production.　And the colony remains economically dependent on the mother-country, both because of its narrow specialization in production and because of its need of protection as against the natives. The failure of the staple crops means ruin, and the natives always outnumber and are generally ready to turn upon their masters.

This sharp contrast in the industrial organization, following upon environmental diversities, is reflected in the secondary societal forms. As I have described rather fully the mores of the frontier society in temperate regions, I shall confine myself in the main to a list of points of difference shown in the other climatic environment.　In place of the small freehold we find here the manor and the chartered company; this is a profound difference in organization, resulting among other things in much absenteeism in tropical colonies, with consequent inevitable results in the treatment of the labor force and in the general societal organization. The tropical colonies, where the natives persist, have also been the classic grounds for a missionary activity unknown in intensity where the immigrants could quickly eliminate the natives.

Women, in the tropics, being for the most

part native consorts, are not valued highly; and the importance of children — who are mainly half-breeds — is less. This reacts upon the whole domestic organization, which cannot be of the type I have sketched as characterizing the temperate frontier. Scarcely anything I have said of this latter type is true of marriage and the family in the tropics.

In the hot lands population is socially as well as biologically heterogeneous, and in time largely mongrel. There is no middle class, but the Europeans are investors and aristocrats or officials at one end of the scale, and reckless adventurers at the other. Slavery flourishes as the only sweeping method of industrial adaptation to life-conditions, but it divides the population sharply and creates widely divergent classes. Wages are low or are absent (under slavery); it is not possible to utilize the land without considerable capital; there is little temptation to aspire to ownership of land.

And then the economic dependence works out into political dependence: tropical dependencies are protectorates and crown colonies. There is, instead of equality and democracy, inequality in many grades and aristocracy. A tropical colony could as readily become a democracy as a temperate colony could become an aristocracy. The destinies of the two are diverse; the one remains under some sort of tutelage, while the other develops at length into an independent state.

Here, then, we have the outlines of a significant example of adaptation in the mores, where those of self-maintenance first adapt themselves to the environment, and then form the basis for a societal code consistent with them. The reader will be able to fill out many details of these outlines. This example likewise throws back light upon the contention as to the basic character of the mores of self-maintenance. And, finally, I think an appreciation of the nature and course of development of frontier communities will make clear to the reader the significance for the science of society of the study of such societies. As in the case of primitive types, their processes of adaptation can be more readily apprehended than those of an older and more complex group; and then their development takes place rapidly through various stages, in a recapitulary manner, and enables us the more easily to derive the origin and nature of the complex from those of the simple.

This is one of the "experiments" which nature performs for the scientific student of human society; it is hardly paralleled on the large scale, it cannot be repeated indefinitely, and so it deserves the more careful examination and analysis.

CHAPTER X

ADAPTATION (*Continued*)

WHEN we come to adaptation in the mores of a highly civilized society the subject reveals at once an extreme complication. For, to use the figure employed before, the life of such a society, in an artificialized environment, is lived on a sort of scaffolding or staging — on a plane above the natural surface. It must always rest upon the latter and be subject to the results of the oscillation or upheaval of its support; but where this sustaining *terra firma* shows roughnesses and inequalities, these are smoothed out, as it were, upon the artificial staging. To change the figure, the road of civilization is like a railway-line; it follows in general the topography of the earth-surface, but it tunnels certain obstacles and spans others, preserving a general level in spite of considerable natural irregularities. The more perfect the roadway, the less does it turn aside and otherwise conform to the minor features of natural environment.

Thus do the mores of a civilized society reflect less faithfully the minor features of the physical environment; they conform to the staging rather than to the supporting surface — to the artificialized environment rather than to the natural. Let us then consider somewhat more closely than we have done hitherto the nature of this artificialized environment. But let us never forget, in so doing, that we have exemplified in it a grand product of adaptation. By learning the laws of nature and conforming to them men are able to construct a railway line passing through mountains and over chasms; and by the same process they have built up the staging upon which civilized people pass their existence, the artificialized environment in which they live and move.

To preserve and upbuild civilization there must be labor and capital, and free interchange of products, material and other. Hence the mores of civilized peoples must include industry, foresight, and mutual toleration; and the opposite qualities of slothfulness, improvidence, and animosity must be represented in the minimum. War, which is a sort of reversion to a primitive form of selection, may tear away whole sections of the civilization-staging;

it often destroys the useless and the unfavorable variations, but it is not discriminating, and much that must be replaced goes down in the general catastrophe. War takes its toll from both labor and capital, the two pillars upon which societal self-maintenance rests; and it handicaps trade. The recognition of this fact — which is periodically verifiable — weakens in the mores the age-long sanction placed on war, and gives carrying-power to the assertions and arguments, however flimsy, advanced by the members of peace societies. Every one reads and speaks of the traits or virtues upon which civilization rests; and these represent, in a broad way, the adaptations in the mores to the artificialized environment.

The form of competition under civilization becomes, as we have seen, industrial rather than military, and thus, as we have also seen,[1] there arises in the society a set of mores consistent all along the line, which Spencer summed up under the term industrialism, as contrasted with militarism. This is perhaps the broadest designation for the congeries of mores and institutions which represent adaptation to the artificialized environment; and a realization of the

[1] P. 150 above.

extent and consistency of development of this type of code under advancing civilization is, perhaps, the strongest element in a conviction that the mores are adaptive, that is, evolutionary. The rise of industrialism, depending as it does on peace and settled order, would have been impossible and, even if possible, a case of maladaptation in uncivilization; it is an adaptation to the artificialized environment, not to the natural, and is paralyzed and ruined where the supports of civilization give way under some catastrophe (earthquake, war) and the society is cast down into the primitive. This is irrespective of the question as to whether the structure may not at length be better built than it was before.

If we try to come down to the more concrete instances of adaptation in the mores of civilized peoples, we can no longer attain clear results by taking a nation or people for an example. Such a social unit comprises too many strata of civilization. The United States, for example, though generally regarded as a nation, contains elements on all stages of culture. This is true in large measure of any unit which could be selected. Civilization is like the snow-line: it crosses all countries and nations horizontally. In a very real sense, the most highly cultured people of all lands have more in common — more mores in common — than do the cultured and the uncultured in any one geographical or ethnological unit. That, under pressure, the sense of unity falls back on the old lines, clinging to the essentials of group-identity and self-maintenance, detracts in no way from the truth of this proposition.

However, we are not seeking for the code of the *literati* of the earth, and are not to be cast down if our unit is somewhat complex. I suppose, in a general way, that one of the chief results of the growth of civilization has been the development of the city. The unparalleled increase in the number of great cities and in the proportion of urban to rural population has constituted a characteristic feature of the last century. Numbers and the contact of numbers form, as we have found,[1] the very conditions of the development of civilization. The modern great city is, physically at least, the most artificialized of environments. If grass and trees are to be found at all in the more densely populated sections, they are sedulously preserved in a sort of museum, the park. I need

[1] P. 21 above.

not enlarge upon this point; a glance at the sky-line of New York City impresses it at once, when one puts aside the indifference of familiarity. And so I propose to use the modern great city as the best example at hand after which to outline the adaptation of the mores to the artificialized environment. To go into much detail would be to rehearse commonplaces known to many. The numerous studies of urban *versus* rural conditions and habitudes can be recalled by the reader, to point by contrast and to amplify the few generalities to be cited here.[1]

Self-maintenance in the city is attained very

[1] Standard works on city conditions furnish an abundance of instances which need not, I think, be quoted here. I have tried to suggest their place in the arrangement of the evidence for evolution in the mores. Such works are: Hurd, R. M., "Principles of City Land Values"; Gemünd, W., "Die Grundlagen zur Besserung der städtischen Wohnverhältnisse"; Weber, A. F., "Growth of Cities in the Nineteenth Century" ("Columbia University Studies," vol. XI, for 1899); Schott, S., "Die grossstädtischen Agglomerationen des Deutschen Reichs," 1871–1910; Pratt, E. E., "Industrial Causes of Congestion of Population in New York City" ("Columbia University Studies," vol. XLIII, for 1911); Rowe, L. S,. "Problems of City Government"; Wilcox, D. F., "Great Cities in America"; Howe, F. C., "European Cities at Work"; Woods, R. A., and others, "The City Wilderness," "Americans in Process"; Strong, J., "The Challenge of the City"; Addams, J., "The Spirit of Youth and City Streets."

largely through an almost unvaried routine, which reduces the ordinary man, as is often remarked, to a cog in the great social machine. It is complained that this is counterselective,[1] for it tends to remove the premium on intellect, initiative, and resourcefulness existing in more primitive conditions. But it is none the less an adaptation, on the plane of societal selection.[2] As in a modern army, the activities are carried on group-wise, and the thinking is specialized; it is the inventor of the machine who does the thinking, and the captain of industry who lays out the campaign. For the competition is between such large bodies of industrial warriors that Indian tactics would be ineffective as maladaptations. This may be regretted, but it is quite inevitable.

Similar mass-organization is visible everywhere in self-maintenance, as, *e.g.*, in housing, transportation, standardization of prices, protection of the market, and so on; and the habitudes of going and coming, eating and sleeping, show a response to this organization which does not need description. Consider the dweller in an apartment, who travels miles every day,

---

[1] Schallmayer, "Vererbung und Auslese im Lebenslauf der Völker," p. 122.   [2] Cf. pp. 177 ff. above.

but who comes into contact with nature only in the most sporadic and distant way; he moves in an artificial environment of materials, sounds, systems, crowding fellow-men, etc., and lives and secures self-realization as he fits into it and is part of it. Thus is developed a group of urban mores which are contrasted, as city ways, with country ways, a contrast which has often been brought out facetiously by picturing the adventures of the rustic in the city. Urban mores are reflected with fidelity in the city newspapers; and the contrast with rural ways is brought out by cartoons and in items from country exchanges which are selected for their incongruity with the urban code.

The mores of the city are also more subject to acculturative influence, by reason of the fact that the city is at the crossways of lines of communication while the country districts remain in isolation. A foreign practice such as tipping secures a hold more quickly in such centers. Hence the city mores are "cosmopolitan." And even apart from acculturation, they are likely to show more variations as numerous and heterogeneous elements come into contact; they are less traditional and consequently less stable, but at the same time offer a busier field for selection. Hence what we call advance or progress is likely to appear in the urban code, where the rural may seem unprogressive, or poorly adaptive. But, on the other hand, variations of the characteristic mode may be carried to such extremes, as in the exaggeration of mass-housing, that the ruder selective processes, operating through under-nourishment, disease, and death, are evoked.

I think it is not necessary to catalogue the cases of adaptation to the artificialized environment as shown in the industrial organization. Labor conditions and those of capital show such adaptation. Wages, prices, and rents reflect the peculiar life-conditions. The property-system develops refinements which are selected automatically or consciously out of numerous variations, as adaptive to the environment. In short, to use a classification employed above, production, consumption, and distribution are of a type unknown before the development of such aggregations of population. Nor do I think it needful to more than direct attention to the mores of societal self-gratification that are characteristic of the city and are in striking contrast with those developed in a different physical and societal

environment. Public amusement is highly organized and differentiated; in fact, for many,
the public place of amusement usurps the function of the home as a center of diversion and
social intercourse. And it is noteworthy that
these mores spread, through imitation, and assume forms characteristic of the environment
of adoption, approaching the city code in differing degrees as the local environment approaches that of the great city. Imitations of
New York life are laughable shams in isolated
rustic communities, where they may be no more
than premature and pretentious in a smaller
city. Reflection over details of the mores of self-
maintenance and self-gratification will afford
unmistakable evidence of their adaptation.

The code shows this adaptation also in what
has to do with marriage and the family. I
have already cited the case of deferred marriage
and the decline of the birth-rate of certain classes
as representing adaptation by the whole society
through its most sensitive part, the part most
sensitive, that is, to considerations centering
about the standard of living. The most highly
equipped show this form of adaptation in a
unique degree. But the location of these, or
their goal, is, in general, the city. And while

statistics about the birth-rate in the big cities as compared with the less artificialized environments may prove little or nothing, it is certain, on general principles, that the former does not favor a rapid natural increase. The characteristic mode of city growth is by accretion.

Living conditions impose everywhere a special type of domestic economy. The women are wage-earners, and so, in some degree, are the children. Regarding the latter, the problem itself of how to raise children in a city is still before the urban community and the adequate adaptation is yet to come. For any but the wealthy, children must be reared in crowded spaces and in contact with influences that would not be deliberately chosen. The childless family has much less difficulty in securing living accommodations and domestic service; and, of course, it is those who have the foresight and a standard of living which they refuse to lower, who adapt themselves to this situation most readily. And if, in the artificialized environment, complicated and costly habits of self-gratification have developed, these enter to round out a standard of living which will not be lowered — a standard that could not exist under closer contact with nature. Such a status of the standard of living must inevitably affect the mores of societal self-perpetuation.

This is a very general relation of adaptation; but I think that if one gives consideration to the sex-mores and those of domestic life, he will find detailed instances of adaptation, called for by the nature of the artificialized environment, occurring to him all the time. The conditions of the association of young men and women which lead to courtship and marriage are largely set for them by the general life-conditions. The qualities which are attractive in the sexes are not always, certainly, the same ones which obtain in less sophisticated communities. What is demanded of the women in marriage is not what the rustic community's mores demand. In short, the pre-conditions of marriage show adaptation of detail, if not always of essence, to general life-conditions of the society.

Satisfactory family life rests upon the same general basis everywhere; but when it comes to the innumerable details, the domestic economy of the city-dwellers has its own peculiarities. Home, as any one knows, is not a spot on the earth's surface, nor anything else that is material; it is an atmosphere. But the character of the environment is not negligible either, and the flat or apartment is not as yet the home that the imagination conjures up nor yet the one that the memory of many recalls. It is often the mere lodging-house and there is no real home at all. Narrow spaces and the proximity of families, again, introduce many an element that makes a special call upon the patience, forbearance, discipline, affection, etc., which go to make up the atmosphere of which I have spoken. Such elements require adaptations in the domestic mores which are, in a sense, parallel to the adaptations called for in community life. And divorce and sex-vice have also their great-city type.

The letter of a religion adapted to conditions of relative uncivilization stands less chance of preservation as it meets an environment more widely differing from that of its origin. The city has the reputation of being more godless than lesser communities. Anachronistic elements are there the more speedily shown to be maladaptive. For example, the life of the urban community depends upon the unintermittent activity of its transportation systems; it could not cease labor for one-seventh of the time and not suffer. Compromise between the exact prescriptions of the religious code and the life-conditions occur all the time, and they represent the adaptation of the former to the latter. Tolerance is the result of insight into the point of view of others; but such insight is to be gained through close association with those whose standpoints are not ours; and this association is most likely to occur where numbers are brought into recurring contact in daily life. Here appears, then, a sort of amalgamation of heterogeneous elements such as results from primitive conquests, with the consequent broadening of the code along all lines, including those of religion. Religion is the great sanction of the mores, and if the latter are altered by adaptation to a typical environment, the former must at length change form to follow them.[1] There is much in the life of a great city which religion in its provincial form could not sanction. But the mores go their way without such sanction; and the alienation of the masses from the established forms of religion follows. The Bible reflects rural conditions, and, unless interpreted as something of wider reach than its own setting, does not fit

[1] Cf. Sumner, on "Religion and the Mores," in "War and Other Essays," Chap. V.

city life and its problems. The power thus to interpret religion is yet undeveloped in the ordinary expositor. Hence the condemnation of the great city for its wickedness, which has issued from many sources during the period of city-development.

The apparatus of control and administration of great cities is yet following afar off upon their rapid material development; adaptation is still very imperfect. But that there is coming into being a system of control which is quite different in degree and complexity from anything the world has yet seen, admits of no doubt. New situations arise so rapidly and suddenly that the existing regulative machinery is always strained beyond its capacity. Much of the time we are thrown back on luck and fall into grave errors; there has not been time to take in the situation and exercise rational selection.

The two great issues in regulation (which are really one) are: the defining of spheres of rights for crowding individuals; and, more inclusively, the safeguarding of the society's life in the highly artificialized environment. Hence, on the one hand, rules for personal behavior that go into a detail uncalled for outside of a densely populated area; and, on the other, broad regulations and far-reaching provisions that scarcely visualize the individual at all. For example, take the traffic-rules of all varieties as a case of limitation of individual freedom of movement; and consider the building and fire codes as broad provisions for the welfare of community life. Everywhere it becomes necessary to resort to legislation, inspection, and enforcement; but everywhere at the same time are to be seen the results of maladaptation. The common interests have multiplied so fast and are so complex — including representatives throughout the whole length of the scale, from small-group interests to class and racial interests — that experimentation and rational selection have not been able to keep pace. Hence adaptation is as yet far from perfect.

Parallels might be cited from other fields. As ocean-liners have become larger and faster, it has been seen through sad experience that the regulations for running them have lagged behind. People have been thinking only of size and speed, and neither builders, owners, crew, nor passengers have really known the new vehicles or the conditions within which they must operate to secure safety. It has been largely an empirical matter. Says Sumner of railroads : [1]

"We have only just reached the point where a few men are competent to manage great lines of railroad on their technical side; we have only just begun to educate men for the railroad business as a profession. Railroad men do not seem yet to have any code of right behavior or right management between themselves — people often deride the professional code of lawyers or doctors, but the value of such a code is seen if we take a case like the one before us, where a new profession has not yet developed a code. The social and economic questions raised by railroads and about railroads are extremely difficult and complicated; we have not, so far, accomplished much of anything toward solving them by experience or theory. The discussion, so far as it has yet gone, has shown only that we have the task yet before us and that, so far, all has been a struggle of various interests to use railroads for their own advantage. The true solution of the only proper legislative problem, *viz.*, how to adjust all the interests so that no one of them can encroach upon the others, has scarcely been furthered at all. It is only necessary to take up a volume of the evidence taken by one of the Congressional committees on this subject, or any debate about it which has arisen in Congress, to see how true it is that conflicting interests are struggling for advantage over each other."

Similarly with a great city; the need of adaptation to this novel environment has been

[1] "The Challenge of Facts and Other Essays," p. 178.

so recent of development that it can as yet be met only in part. Old methods will not do; instance the insufficiency of rural charity organization in the big city. Naturally a passable adaptation is first achieved along the line of the self-maintenance mores; elsewhere in the societal field variation and experiment are rife and the most strenuous effort is being put forth to understand, and then to perform a successful selection. In any case the code comes to include new elements and, even in its inadequacy and incompleteness, grows characteristic.

The immigrant peasant finds the code of personal conduct oppressive and incomprehensible, and wonders if the eulogies of America as the land of freedom are not overdone. It is wrong to peddle wares freely where one chooses, to control the time and activities of one's own children, to make such noises as one wishes, even to expectorate when and where one pleases. However, a species of half-adaptation, or adaptation by indirection, comes about, if the exponent of folkways alien to the code obtains a buffer in the form of a patron — ward-boss or other — who will for a consideration secure exemption or evasion for the culprit. And, indeed, where there are so many who adhere to

alien codes, as usual the prescriptions and
taboos have to be enforced indulgently for a
time. In general, regulations aim at practice
which are normal in the rural environment
but are both survivalistic and inexpedient in
that of the big modern city. European peas
ants may try to celebrate their harvest festival
in a thin and shadowy form; but they may no
try to keep goats, chickens, and pigs in tene
ments. Gradually such survivals are elimi
nated, especially by the education of the nex
generation. But they are not wanting even in
the regulative system itself; in American city
government there is a persistence of the rura
forms, giving opportunities for "graft," the
"spoils system," and general inefficiency. The
need of reform in the regulative organization
of great cities is a topic of constant agitation
Says Mr. McAneny,[1] an official whose extended
service in, and study of city government give
him a right to speak: "The plan gaining in
favor in both States and cities is to follow the
example of the nation, to confine the list of
elective offices to those chosen for the broade
business of government, and to concentrate in
a single executive the responsibility for the ap

[1] *New York Times*, May 17, 1914.

pointment of the rest. The establishment of the 'short ballot' means nothing more or less than this. It is of the essence of common sense. To expect the voter, however high his intelligence, to choose with proper discrimination between a multitude of candidates for minor positions, when he has heard little or nothing about either the candidates or the positions, is to expect the impossible. Under cloak of this system, political organizations have been permitted for years to run in their unfit men with the fit, and thus capture offices that, though inconspicuous, are profitable from the political point of view. It is a necessary part of the complete reform of the electoral machinery of cities that an end be put to this practice."

Not all these regulations are peculiar to the great city, but they arise in response to the need of adaptation to a crowded and artificialized environment of which the great city is the extreme type. More broadly, the community interest demands that great enterprises in further developing the artificiality of the environment be carried out (sewage-systems, water-supply, rapid-transit), and that the society be fitted to exercise rational selection in its further adaptation (school-system, election-reforms). And so an enormous and complicated system of societal control evolves in response to the society's life-conditions, for which simpler aggregations felt little more need than does the Papuan for a clearing-house.

This trio of illustrations of adaptation in the mores — and in the institutions that crystallize out of the mores — on the part of three very differently situated human societies, seems to me sufficient at least to suggest many other and diverse cases of the same process. But such adaptation is the end-result of evolution. If we can accept the conclusion, stated in advance of the evidence at the outset of chapter VIII, that every established and settled human institution is justifiable, in its setting, as an adaptation, it seems to me that we are thereby accepting the extension of the Darwinian theory to the field of the science of society. But the attempt has been made in the foregoing to show also that this end-result has been reached through the operation of factors essentially the same as those whose operation in the organic field leads to organic adaptation. It is freely admitted that we cannot demon-

strate processes and results in the societal field
with the precision and under the authority of
verification which are characteristic of the natural sciences; societal evolution is on a different plane, and the social sciences, in their different mode, have as yet no such equipment of
evidence and method as have the natural
sciences. "Aller Anfang ist schwer," says
Goethe; and we might venture to modify, un-
rhythmically, the rest of his line: "vor allem
der Anfang der Wissenschaft." But one of
the difficulties of the social sciences is, as I said
at the beginning, that they have no evolutionary
orientation such as Darwin gave to the natural
sciences. The question was: Cannot this
orientation be given to them by extending the
evolutionary theory to the field which they
cover? This cannot be done, to any purpose,
by "reasoning from analogy." I have come
to believe that there is more here than analogy,
and have tried to convey some of the reasons
for that conviction. If what I have written
helps any one else to arrive at the same conviction, I think he will find it useful in securing an
orientation for the study of human society.

NOTE

IT will perhaps occur to some close student of Sumner's work that Sumner himself has written on "Evolution and the Mores" (see "Folkways," preface, *ad fin.*). This is correct; and I feel that I must say something about that essay in this place. Some time before the publication of "Folkways," Professor Sumner read, before the Anthropology Club of Yale University, a paper entitled, "The Application of the Notions of Evolution and Progress on the Superorganic Domain," in which he set forth his conviction that there is no progressive evolution in the folkways. This paper summarized, to some extent, a much longer manuscript on "Sociology and Evolution," the last passages of which I have quoted at the beginning of Chapter VIII above. I think this was the projected chapter of "Folkways," mentioned in the preface to that book.

At the club meeting which I have mentioned there were present representatives of several other departments — biology, geology, psychology, history — and there ensued upon Sumner's paper a prolonged discussion from the different points of view represented. Several men then undertook to set forth, at succeeding meetings, their views on the subject. Sumner was present on all these occasions and listened to the criticism with his usual care. He also fought back lustily, as was his wont; but as we returned one evening from the last of these discussions, he told me

328

he had decided not to include the chapter on evolution and the mores in his forthcoming book. I judge from the unfinished character of the manuscript that he laid it aside at that time and did not return to it again. Other matters occupied his mind during the last few years of his life; in the course of many conversations I do not recall any further reference to the topic.

This essay of his, as it seems to me, would be relevant and compelling if he could have completed it under the title "Progress and the Mores," or with the understanding that "evolution" connoted the process as Spencer saw it. Criticism of Sumner's utterance was confined chiefly to the position taken that evolution and progress are synonymous, or nearly so. Sumner admitted that he meant "progressive evolution" and endeavored to show that he was correct in that view. I expect sometime to utilize the bulk of this essay in connection with other unfinished manuscripts left by him. As a refutation of much foolish talk about progress, some of its trenchant paragraphs should be very effective.

As for the systematic application to the folkways of the central idea of Darwinian evolution — adaptation to environment, secured through the operation of variation, selection, and transmission — I do not believe that it occurred to Sumner to undertake it. Often in "Folkways," and even in much earlier writings, he uses some of the terms, as, for instance, variation and selection ("Folkways," §§ 88, 170, *et al.*); and frequently the operation of factors, as developed above, is implied; but that was as far as he went. What he wanted to make clear was the origin and nature of the folkways; and then he meant to hasten back to his "Science of Society," rewrite in the light of "Folkways" what he had already written, and complete the treatise.

It is my belief that he would have been obliged to return to the topic of evolution in its relation to the folkways before he could have satisfied himself to go on with his "Science of Society." No one could be more in sympathy with his general way of looking at the science than I am; but I could not accept his views about evolution and the mores — views which were somewhat unsettled, I thought, by the discussions I have mentioned; and I have been convinced that some understanding must be arrived at respecting societal evolution before it is possible to complete a general book on the science of society that shall rest upon the conception of the folkways. And so I have worked out that which precedes.

This is what lies behind the statement in the preface that the present volume is an extension upon the idea of the folkways.

INDEX

A

Acculturation, 216 ff., 258, 263, 277, 313.

Acquired Characteristics, 209–211.

Adaptation, 9, 23, 41, 92, 129, 154, 160, 164, 187, 215, 216, 227, 234, 245–247 ff. *See* Darwinian Factors.
  Mental and Social, 18 ff., 21 ff.
  Physical *vs.* Mental, 18.
  To Geographic Environment. *See* Anthropogeography.

Adat, 155.

Advertising, 244.

Aged, 60.

Aleatory Element, 74.

Amalgamation, 46, 75 ff., 80, 242–243, 319.

Ambel anak, 155.

American Traditions, 79.

Anachronism. *See* Maladaptation.

Analogy, 4, 14–15, 46, 209, 220–222, 327.

Animism *See* Religion.

Annihilation, 64 ff , 72, 135.

Antagonistic Coöperation, 192.

Anthropogeography, 256 ff.

Anthropomorphism, 260.

Aristocracy, 115, 148, 291–293, 303.

Artificialized Environment, 67, 306 ff

Artificialization, 71, 177 ff.

Arts of Life, 23, 138, 248, 279. *See* Maintenance-mores.

B

Assimilation. *See* Amalgamation.

Association, 68.

Austerity, 255.

Australia, 297.

Australians, 156.

Bagehot, W., 11.

Bergson, H , 48.

Biogenetic Law, 220, 225.

Biological Qualities. *See* Selection, Counterselection.

Black Death, 280.

Bland Silver Bill, 113.

Boers, 289.

Borup, George, 271.

Brain, 18, 19.

Bricks, 49.

Bride-price, 155–156.

Bryce, J., 243.

C

Campbell, H., 210.

Cannibalism, 45.

Capital, 146, 182, 286, 291–292, 303, 307, 314.

Capital Punishment, 65.

Catastrophic Theory, 259.

Cause, Single and Multiple, 259.

Celibacy, 173–174, 187.

Chance. *See* Luck.

Chapin, F. S., 11.

Charity. *See* Humanitarianism.

Children, 29, 221, 222, 228, 289, 299, 316. *See* Eskimo.

THE following pages contain advertisements of books by the same author or on kindred subjects.

rinciples of Western Civilization

By BENJAMIN KIDD

Cloth, 8vo, $2.0

FROM THE PREFACE

" The book of which another edition is offered to the reade
as been in one important respect a new departure. It ha
epresented, so far as the author is aware, the first attempt i
estern literature to present a type of civilization as a develop
g system of life possessing a characteristic meaning of its ow
the evolutionary process, and having in this sense an organi
ity far deeper than that of any of the nations or politica
tates included in it.

.   .   .   .   .

" If there is one idea more than another which is to be clearl
arried away from this book, and particularly from that whic
receded it ('Social Evolution'), it is that civilization is not
atter of race, nor descent, nor of superior intellectual capacity
ut of ethos — that kind of ethos which is described in thes
hapters as making of our Western civilization a living, organic
eveloping unity."

" A book which every thoughtful person will have to read, and
hat is more, will wish to read. . . .   Mr. Kidd is nothing i
ot the founder of something like a new school. . . .   In an ag
f apparently increasing Materialism, and with the aid of th
ery calculus which Materialism has been supposed to suppl
nd support, he rehabilitates Idealism, and tells us that in some
hing barely apprehended by our consciousness, beyond the pres
nt horizon and scheme of things, lies the secret, in the long run
ven of material success . . . important in the variety of poin
ouched, the novelty and breadth of the hypothesis and its ap
lication.   It is no less than a new Philosophy of History. . .
ll minor blemishes are of little importance compared with th
and sweep of the whole, which are irresistible.   If the form
ation halts, the general argument develops itself with great an
owing force ; if here and there the writing is inadequate, th
eneral eloquence is very marked, and kindles again and agai
to a glow of beauty and intensity." — *The Spectator* (London)

By EDWARD A. ROSS

Professor of Sociology, University of Wisconsin
Formerly of the University of Nebraska

Social Psychology

Cloth, 8vo, $1

" The volume marks off for itself a very definite field of research, ; scours the circumscribed area in the most scientific manner. . . . I fessor Ross has laid bare the more vital social traits, good and bad, of human mind in a manner calculated to awaken thought." — *The 1 York Tribune.*

Social Control

A Survey of the Foundations of Order

PART I — THE GROUNDS OF CONTROL
PART II — THE MEANS OF CONTROL
PART III — THE SYSTEM OF CONTROL.

Cloth, 12mo, leather back, $1

The author seeks to determine how far the order we see about us is (to influences that reach men and women from without, that is, social fluences. His thesis is that the personality freely unfolding under con tions of healthy fellowship may arrive at a goodness all its own, and t' order is explained partly by this tendency in human nature and partly the influence of social surroundings. The author's task, therefore, is f to separate the individual's contribution to social order from that of s(ety, and second, to bring to light everything that is contained in this cial contribution.

Foundations of Sociology

The Scope and Task of Sociology — The Sociological Frontier of E nomics — Social Laws — The Unit of Investigation in Sociology — M mind — The Properties of Group-Units — The Social Forces — The F tors of Social Change — Recent Tendencies in Sociology — The Causes Race Superiority — The Value Rank of the American People.

The author lays a foundation for a social science that shall withstand (severest tests, and formulates the principal truths about society. In (belief that the time is past for one-sided interpretations of society, the fort is made to group together and harmonize the valuable results of the schools.

Both the above are in " The Citizen's Library " of Economics, Politi and Sociology, edited by RICHARD T. ELY, Ph.D., LL.D., Professor Political Economy, University of Wisconsin.

An Introduction to the Study of Heredity

By HERBERT EUGENE WALTER

Assistant Professor of Biology in Brown University

Cloth, 12mo, 264 pages, with 72 figures and diagrams, $

An excellent summary of the more recent phases of the questions of he
ity which are at present agitating the biological world. The author
unerringly to the fundamentals of our most recent advances, yet the wor
so simple in language and so plain in illustration that anyone intereste
either animal or plant breeding can read it to advantage. It is not or
good text to put into the hands of students taking courses in breeding, he
ity, genetics, eugenics, or evolution, but is also used as a required supplemen
text in introductory courses on general biology or zoölogy.

The First Principles of Evolution

By S. HERBERT

Cloth, 8vo, 346 pages, containing 90 illustrations and tables, $

Though there are hosts of books dealing with Evolution, they are ei
too compendious and specialized, or, if intended for the average reader,
limited in their treatment of the subject. In a simple, yet scientific, man
the author here presents the problem of Evolution comprehensively in al
aspects.

CONTENTS

INTRODUCTION — Evolution in General.
SECTION I.— Inorganic Evolution
    The Evolution of Matter.
SECTION II — Organic Evolution.
PART I — The Facts of Evolution.
    Morphology.
    Embryology.
    Classification.

Palæontology.
Geographical Distribution.
PART II — Theories of Evolution.
SECTION III — Superorganic Evolution.
    Social Evolution
CONCLUSION — The Formula of Evolutior
    The Philosophy of Chang

The Meaning of Evolution

By SAMUEL C. SCHMUCKER

Cloth, 12mo, 298 pages, illustrated, $

What is our origin ?
What are we to be ?
Can we help the great advance ?
Professor Schmucker in this book sets about to meet such questions.
ground of the evolutionary idea itself developed by Charles Darwin, its
ification by later students and the present state of the question, are tol
language wholly devoid of technical terms. The work is clear, compre
sive, but not detailed, and entirely reverent.

OF

Economics, Politics and Sociology

EDITED BY RICHARD T. ELY, PH.D., LL.D.
Of the University of Wisconsin

Each volume 12mo, half leather,

" This has already proven itself one of the most fruitful amor
the different ' libraries ' of the sort, in yielding stimulatir
books upon the modern phases of economic and soci
science." — *Philadelphia Public Ledger.*

CITIZENS' LIBRARY OF ECONOMICS, POLITICS AND
CIOLOGY (Continued).

Introduction to Business Organization. By S. E. Sparling.

Introduction to the Study of Agricultural Economics. By H. C. T

Irrigation Institutions: A Discussion of the Growth of Irrigated
culture in the Arid West. By E. Mead.

Money: A Study of the Theory of the Medium of Exchange.
David Kinley.

Monopolies and Trusts. By R. T. Ely.

Municipal Engineering and Sanitation. By M. N. Baker.

Newer Ideals of Peace. By Jane Addams.

Principles of Anthropology and Sociology, The, in their Relatio
Criminal Procedure. By M. Parmelee.

Railway Legislation in the United States. By B. H. Meyer.

Social Control: A Survey of the Foundation of Order. By E. A.

Some Ethical Gains Through Legislation. By Mrs. Florence K

Spirit of American Government, The. By J. A. Smith.

Studies in the Evolution of Industrial Society. By R. T. Ely.

Wage-Earning Women. By Annie M. MacLean.

World Politics. By Paul S. Reinsch.

NEW SERIES

Cloth, 12mo,

Progressive Movement, The. By B. P. DeWitt.

Wealth and Income of the People of the United States. By W. I.

American Municipal Progress. By Charles Zueblin.

Social Evolution. By Benjamin Kidd.

PUBLISHED BY

THE MACMILLAN COMPANY

Publishers   64-66 Fifth Avenue   New York

CPSIA information can be obtained
at www.ICGtesting.com
Printed in the USA
LVOW13s1955190817
545631LV00019B/1129/P